MATHEMATICS FOR ENGINEERS AND SCIENCE LABS USING MAXIMA

Seifedine Kadry
Pauly Awad

AAP | APPLE ACADEMIC PRESS

Apple Academic Press Inc.
3333 Mistwell Crescent
Oakville, ON L6L 0A2, Canada

Apple Academic Press Inc.
1265 Goldenrod Circle NE
Palm Bay, Florida 32905, USA

Exclusive worldwide distribution by CRC Press, a member of Taylor & Francis Group

International Standard Book Number-13: 978-1-77188-727-4 (hbk)
International Standard Book Number-13: 978-0-42946-971-8 (eBook)

Library and Archives Canada Cataloguing in Publication

Title: Mathematics for engineers and science labs using Maxima / Seifedine Kadry, PhD, Pauly Awad.

Names: Kadry, Seifedine, 1977- author. | Awad, Pauly, author.

Description: Includes bibliographical references and index.

Identifiers: Canadiana (print) 20189068426 | Canadiana (ebook) 20189068434 | ISBN 9781771887274 (hardcover) | ISBN 9780429469718 (PDF)

Subjects: LCSH: Engineering mathematics—Data processing.

Classification: LCC TA345 .K33 2019 | DDC 620.001/51—dc23

Library of Congress Cataloging-in-Publication Data

Names: Kadry, Seifedine, 1977- author. | Awad, Pauly, author.
Title: Mathematics for engineers and science labs using Maxima / Seifedine Kadry, PhD, Pauly Awad.
Description: Toronto : Apple Academic Press, 2019. | Includes bibliographical references and index.
Identifiers: LCCN 2018058079 (print) | LCCN 2018059264 (ebook) | ISBN 9780429469718 (ebook) | ISBN 9781771887274 (hardcover : alk. paper)
Subjects: LCSH: Mathematics--Data processing. | Engineering mathematics--Data processing.
Classification: LCC QA76.95 (ebook) | LCC QA76.95 .S45 2019 (print) | DDC 620.001/51--dc23
LC record available at https://lccn.loc.gov/2018058079

Apple Academic Press also publishes its books in a variety of electronic formats. Some content that appears in print may not be available in electronic format. For information about Apple Academic Press products, visit our website at **www.appleacademicpress.com** and the CRC Press website at **www.crcpress.com**

ABOUT THE AUTHORS

Seifedine Kadry, PhD

Seifedine Kadry, PhD, is currently working as an Associate Professor at Beirut Arab University, Faculty of Sciences, Department of Mathematics and Computer Science, Beirut, Lebanon. He serves as an Editor-in-Chief for the *Research Journal of Mathematics and Statistics* and the *ARPN Journal of Systems and Software*. He worked as a Head of Software Support and Analysis Unit of First National Bank where he designed and implemented the data warehouse and business intelligence. He has published several books and is the author of more than 50 papers on Applied Math, Computer Science, and Stochastic Systems in peer-reviewed journals. At present, his research focuses on system prognostics, stochastic systems, and probability and reliability analysis. He received a PhD in computational and applied mathematics in 2007 from the Blaise Pascal University (Clermont-II) – Clermont-Ferrand in France.

Pauly Awad, MSc

Pauly Awad is currently affiliated with the American University of the Middle East, Kuwait, where she teaches introduction to mathematics and calculus. Formerly, she taught mathematics at the College St. Francois Des Peres Capucins in Beirut, Lebanon. She has a BS degree in Computer Science and an MS degree in Mathematics Education from Lebanese University.

CONTENTS

ABBREVIATIONS

GCD	greatest common divisor
LCM	least common multiple
ODE	ordinary differential equation

PREFACE

This book is designed to accompany any textbook that covers the topics of pre-calculus, calculus, linear algebra, differential equations, and probability and statistics. While these textbooks focus mainly on solving mathematics problems using the paper-and-pencil method, this book teaches you how to solve these problems using Maxima open-source software. Maxima is a system for the manipulation of symbolic and numerical expressions, including differentiation, integration, Taylor series, Laplace transforms, ordinary differential equations, systems of linear equations, polynomials, sets, lists, vectors, and matrices. One of the benefits of using Maxima to solve mathematics problems is the immediacy with which it produces answers. You can rest assured that the time you invest in learning Maxima now will pay off in the future; particularly, if you are a mathematics, science, or engineering student. You will be able to apply nearly all of the Maxima skills that you will learn in this book to your future courses and research.

CHAPTER 1

INTRODUCTION TO wxMAXIMA

Maxima (http://maxima.sourceforge.net/) is an algebra software for the handling of symbolic and numerical expressions, including derivative and integration of functions, series of Taylor, the transformation of Laplace, solving systems of linear equations, solving ordinary differential equations, working with vectors and matrices, the plot in 2D and 3D, and others. Maxima provides high precision and accurate numerical results by using exact fractions variable-precision floating-point numbers. Maxima is an open source software, compatible with Windows, Linux, and MacOS X. Maxima is a descendant of Macsyma, the legendary computer algebra system developed in the late 1960s at the Massachusetts Institute of Technology. Macsyma was revolutionary in its day, and many later systems, such as Maple and Mathematica, were inspired by it.

- **wxMAXIMA vs XMAXIMA**

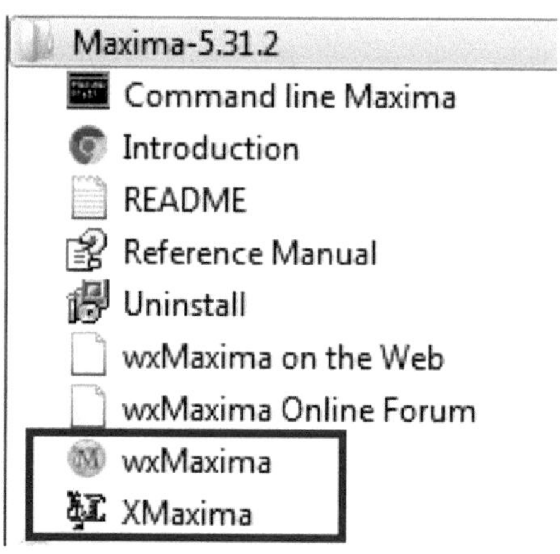

There are two possible instances of *Maxima,* called *wxMaxima* and *XMaxima,* both allow the user access to the Maxima commands, the difference is in their graphical display as shown in the figures below.

1. wxMAXIMA

2. XMAXIMA

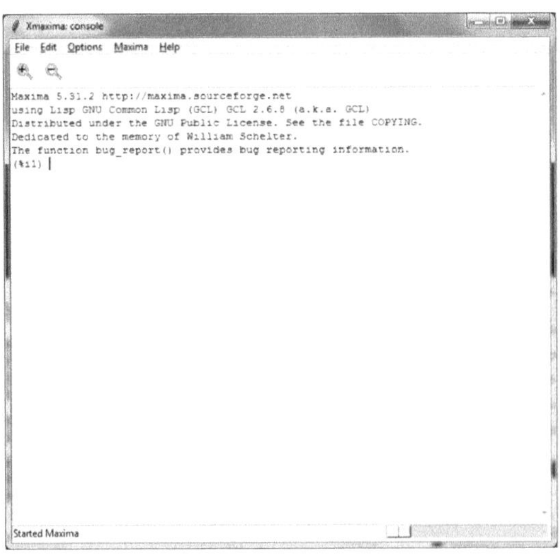

The graphical interface of *wxMaxima* is more refined than that of *XMaxima* because:

- It allows mixing text with mathematical expressions to produce printable documents.
- Some commands can be activated by using the buttons shown at the bottom of the interface, e.g., *Simplify, Factor,* etc.
- It produces true two-dimensional mathematical output.
- It provides dialogues to enter parameters of selected commands.
- It maintains a command line history buffer where previously used commands can be accessed, repeated, or edited.
- It provides many *Maxima* commands in menus.

1.1 wxMAXIMA MAIN MENU

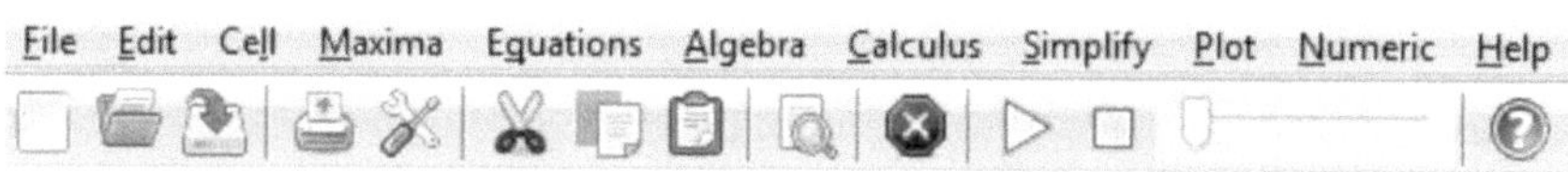

The first row is the main menu in the *wxMaxima* includes:

- The *File* option contains traditional items (New, Open, Save…) and some specific items like *Load package*, *Batch file*, and *Monitor File*.
- The *Edit* option includes items (Find, Copy, Paste…), as well as specific items like Copy as latex…
- The *Cell* option includes specific-items like Evaluate Cell(s)
- The *Maxima* option includes specific-items like restart Maxima, Panes…
- The *Equations* option includes items related to the equation like Solve, Find root, Solve linear system, Solve ODE…
- The *Algebra* option includes specific-items for linear algebra like matrices…
- The *Calculus* option includes specific-items for calculus like derivation, integration…
- The *Simplify* option includes specific-items like expand, contract…
- The *Plot* option includes specific items for a plot in 2D and 3D
- The *Numeric* option includes items for formatting the display.
- The *Help* option contains several items like:
  - *Maxima help*: opens the *Maxima Manual* window with description and examples of *Maxima* commands.

- *Describe*: produces a dialogue where the user can enter the name of a specific command.
- *Example*: shows a series of examples of applications.
- *Apropos*: to search for a keyword
- *Show tip*: shows tips on the use of *Maxima*.
- *About*: provides the current version of *wxMaxima*.

1.2 WXMAXIMA TOOL BAR

The second row is the toolbar in the *wxMaxima* includes:

(1) (2) (3) (4) (5) (6) (7) (8) (9) (10) (11) (12) (13)

(1) To create a new document
(2) To open existing document
(3) To save a document
(4) To print a document
(5) To configure *wxMaxima*
(6) Cut
(7) Copy
(8) Paste
(9) Find and replace
(10) Interrupt current computation
(11) Start animation
(12) Stop animation
(13) Show Maxima Help

1.3 USING THE INPUT LINE

The *INPUT* line is the white window under the toolbar. This window can be used to:

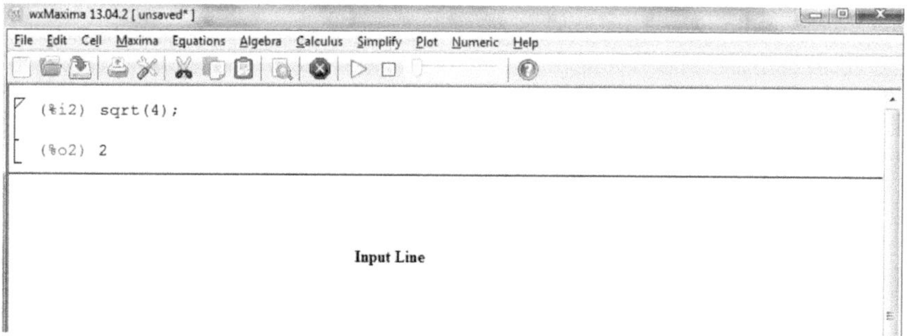

a. Perform a calculation like:

```
(%i1) sqrt(1+3.5^2)/sin(%pi/6);
(%o1) 7.280109889280518
```

b. Define variables like:

```
(%i2) a:2;

(%o2) 2
```

c. Define a function like:

```
(%i3) f(x):=sqrt(1+x^2);
```

(%o3) $f(x):=\sqrt{1+x^2}$

d. Evaluate a function like:

```
(%i3) f(x):=sqrt(1+x^2);
```

(%o3) $f(x):=\sqrt{1+x^2}$

```
(%i4) f(2/3);
```

(%o4) $\frac{\sqrt{13}}{3}$

e. Plot a function like:

```
(%i5) plot2d(f(x),[x,-2,2]);
```

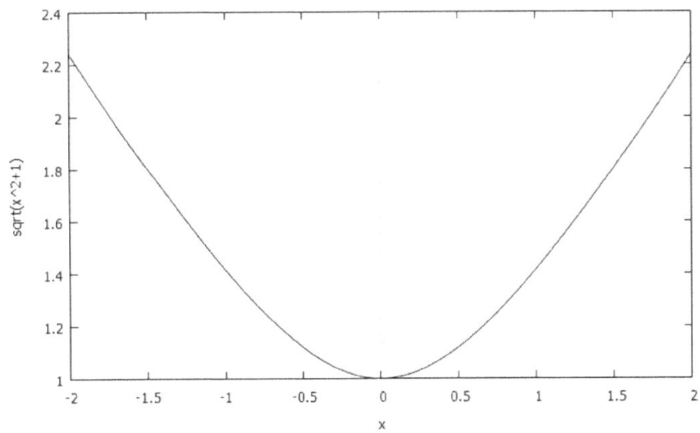

f. Enter other types of functions or operations like derivative:

```
(%i1) diff(t^2*sin(t), t);
```

(%o1) $2\,t\,\sin(t)+t^2\cos(t)$

Below some rules of syntax to follow while using the input line:

- Use a colon (:) to assign the value of a variable
- Use a colon followed by the equal sign (:=) to define a function
- *Maxima* expressions end with a semi-colon. If you forget to enter the semi-colon in the *Input* line, *wxMaxima* will enter it for you.
- The name of a Variable or function must start with a letter
- The following words are reserved to *Maxima* and cannot be used as variable names: integrate, next, from, diff, in, at, limit, sum, for, and, else if, then, else, do, or, if, unless, product, while, thru, step.
- The following functions are predefined in *Maxima* and cannot be used as variable or function names:

sin	sine	*sqrt*	square root	*cos*	cosine
cot	cotangent	*tan*	tangent	*sec*	secant
asin	inverse sine	*csc*	cosecant	*acos*	inverse cosine
acot	inverse cotangent	*atan*	inverse tangent	*asec*	inverse secant
exp	exponential	*acsc*	inverse cosecant	*log*	natural logarithm
cosh	hyperbolic cosine	*sinh*	hyperbolic sine	*tanh*	hyperbolic tangent
acosh	hyperbolic acos	*asinh*	hyperbolic asin	*atanh*	hyperbolic atan
ceiling	integer above	*floor*	integer below	*fix*	integer part
		float	conver to floating point	*abs*	absolute value

NB: by default, *Maxima* display symbolically the results, i.e., including fractions, square roots…, instead of floating-point results. We can use the function *float,* to convert the result to floating-point solutions.

- To refer to the immediately preceding result computed by Maxima, we can either use its o label, or the special symbol percent (%). Here is an example of six consecutive inputs then six outputs.

```
(%i2) exp(-2.5)*sin(3*%pi/11);float(%);exp(-3);float(%);log(5);float(%);
```

(%o2) $0.082084998623899\,\sin\left(\frac{3\,\pi}{11}\right)$

```
(%o3) 0.062035702770881
```

(%o4) $\%e^{-3}$

```
(%o5) 0.049787068367864
(%o6) log(5)
(%o7) 1.6094379124341
```

- Below is a list of a mathematical constant available in *Maxima*:

`%e`	base of the common logarithms (=`exp(1)`)
`%i`	imaginary unit (=`sqrt(-1)`)
`inf`	real positive infinity
`minf`	real negative infinity
`infinite`	complex infinity
`% phi`	the golden ratio (ϕ)
`% pi`	ratio of length of circumference to its diameter (π)
`%gamma`	Euler's constant (γ)
`false, true`	boolean values (or logical values)

- Imaginary or Complex numbers in *Maxima*: The imaginary number *i* is entered as % i in *Maxima*. Examples:

```
(%i1) z1:3+5*%i;
(%o1) 5 %i +3
```

Defining the complex number z1

```
(%i2) z2:-2+6*%i;
(%o2) 6 %i -2
```

Defining the complex number z2

```
(%i3) z1+z2;
(%o3) 11 %i +1
```

Adding z1 and z2

```
(%i4) z1-z2;
(%o4) 5 - %i
```

Subtracting z1 and z2

```
(%i5) expand(z1*z2);
(%o5) 8 %i -36
```

z1 * z2

```
(%i6) expand(z1^2);
(%o6) 30 %i -16
```

z1 to the power 2

- Predefined functions of a complex number in *Maxima*:

cabs	(complex *abs*olute value) calculates the modulus
carg	(complex *arg*ument) calculates the argument
rectform	generate rectangular (Cartesian) form
polarform	generate polar form
realpart	extract the real part
imagpart	extract the imaginary part
conjugate	calculates the complex conjugate

Examples:

```
(%i7) cabs(z1);
(%o7) √34

(%i8) arg(z1);
(%o8) arg(5 %i +3)

(%i9) z2;-z2;
(%o9) 6 %i -2
(%o10) 2 -6 %i

(%i11) conjugate(z2);
(%o11) -6 %i -2

(%i12) expand(z2*conjugate(z2));
(%o12) 40
```

(%i13) `rectform(z1/z2);`

(%o13) $\frac{3}{5}-\frac{7\,\%i}{10}$

(%i14) `rectform(sqrt(z1));`

(%o14) $\frac{\sqrt{\sqrt{34}-3}\,\%i}{\sqrt{2}}+\frac{\sqrt{\sqrt{34}+3}}{\sqrt{2}}$

(%i15) `polarform(z1);`

(%o15) $\sqrt{34}\,\%e^{\%i\,\mathrm{atan}\left(\frac{5}{3}\right)}$

(%i16) `polarform(z2);`

(%o16) $2\sqrt{10}\,\%e^{\%i\,(\pi-\mathrm{atan}(3))}$

1.4 USING THE BUTTON PANEL

There are 18 buttons in the *xwMaxima* General Math menu (in Maxima option then Panes) these buttons can be used for common operations. For example:

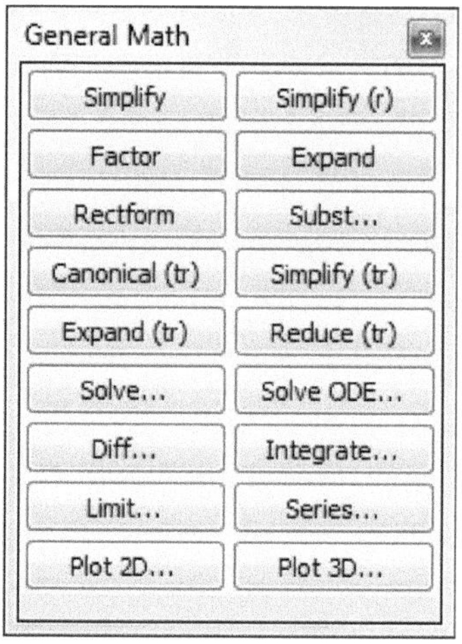

The operation of the buttons, with appropriate examples, is shown next.

Simplify: to simplify expressions. We use this button after an output expression otherwise we get incorrect syntax:

```
(%i17) (x+2)*(x-2);
(%o17) (x-2)(x+2)

(%i18) ratsimp(%);
(%o18) x^2-4
```

Simplify(r): simplifies expressions containing logs, exponentials, and radicals:

```
-->    (%e^x-1)/(%e^(x/2)+1);

(%i19) radcan((%e^x-1)/(%e^(x/2)+1));
(%o19) %e^(x/2)-1
```

Factor: factors an algebraic expression:

```
-->    x^2+y^2-2*x*y;

(%i22) factor(x^2+y^2-2*x*y);
(%o22) (y-x)^2
```

Expand: expands an algebraic expression:

```
-->    (x+1)*(x-1)*(x^2+1);

(%i23) expand((x+1)*(x-1)*(x^2+1));
(%o23) x^4-1
```

Solve: solves an equation:

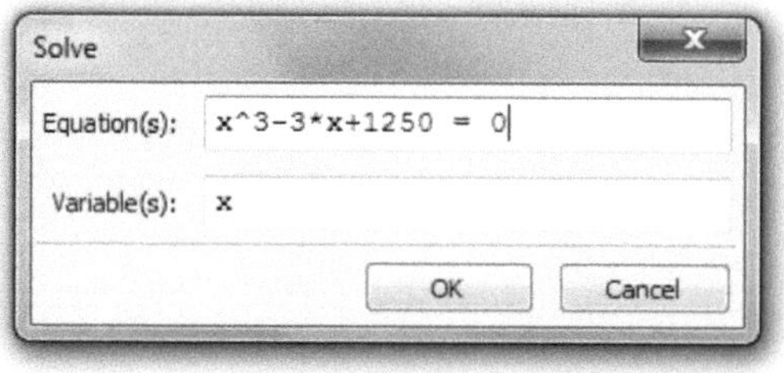

(%i1) solve([x^3-3*x+1250 = 0], [x]);

(%o1) $[x=\dfrac{\frac{\sqrt{3}\,\%i}{2}-\frac{1}{2}}{(4\sqrt{24414}-625)^{1/3}}+(4\sqrt{24414}-625)^{1/3}\left(-\frac{\sqrt{3}\,\%i}{2}-\frac{1}{2}\right), x=(4\sqrt{24414}-625)^{1/3}\left(\frac{\sqrt{3}\,\%i}{2}-\frac{1}{2}\right)+\dfrac{-\frac{\sqrt{3}\,\%i}{2}-\frac{1}{2}}{(4\sqrt{24414}-625)^{1/3}}, x=(4\sqrt{24414}-625)^{1/3}+\dfrac{1}{(4\sqrt{24414}-625)^{1/3}}]$

Plot 2D: produces an *x-y* (two dimensional) plot:

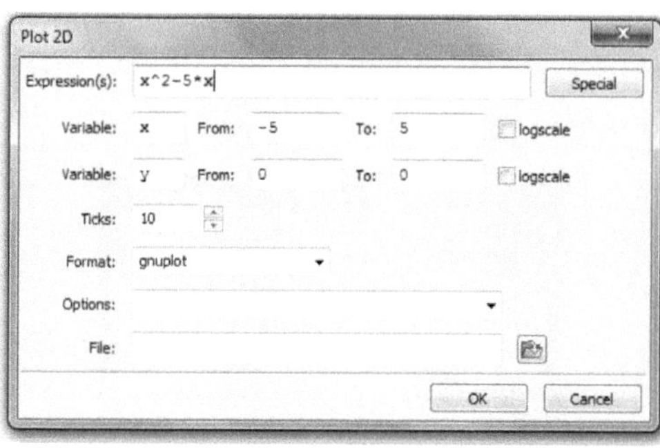

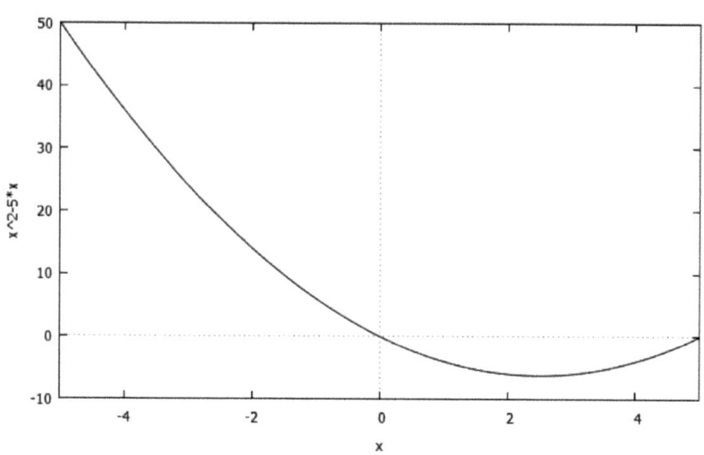

Simplify(tr): perform trigonometric simplification in terms of *sin* and *cos*.

Expand(tr): expands a trigonometric expression like cos(x+y).

Reduce(tr): convert powers of trigonometric functions to those of multiples of the angle.

Rect form: produces the rectangular form of a complex number.

Solve ODE: solves a 1st order or 2nd order ordinary differential equation like:

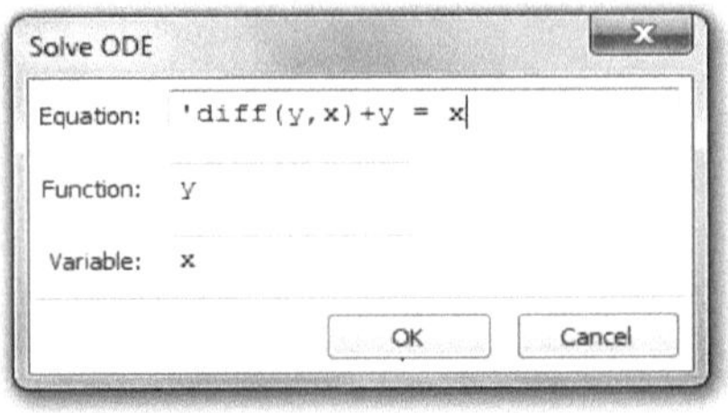


```
(%i5) ode2('diff(y,x)+y=x, y, x);
```

(%o5) $y = \%e^{-x}\left((x-1)\%e^{x} + \%c\right)$

Plot3D: plot a three-dimensional function like:

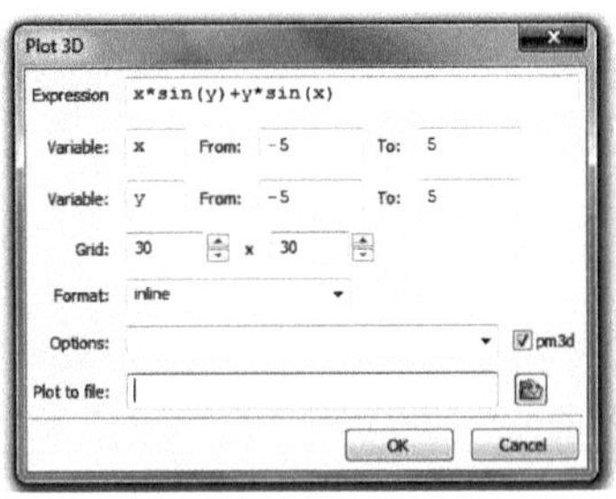


```
(%i6) wxplot3d(x*sin(y)+y*sin(x), [x,-5,5], [y,-5,5])$
```

(%t6)

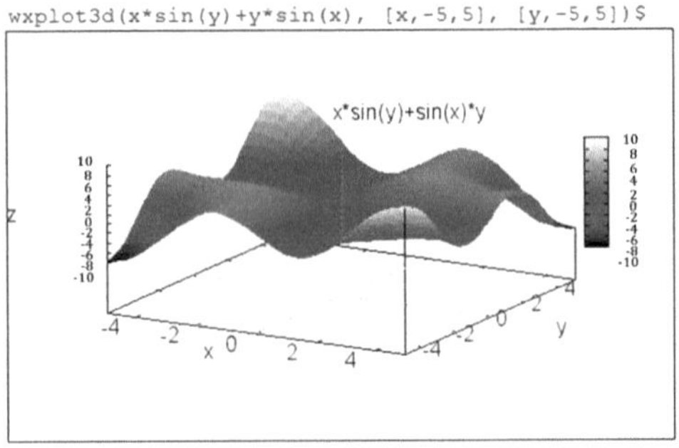

(See color insert.)

subset: substitute an expression with a new variable like:

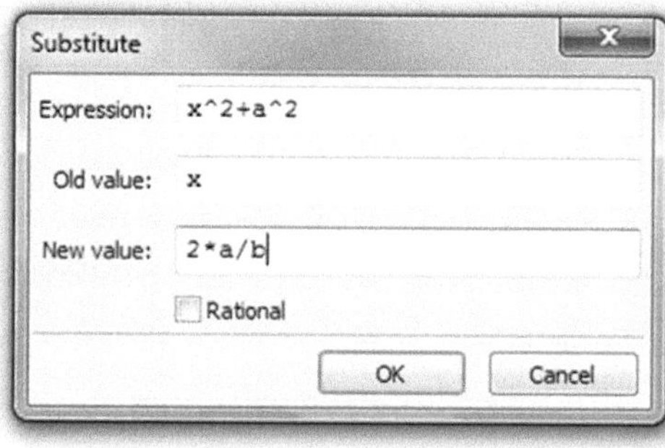


```
(%i13) subst(2*a/b, x, x^2+a^2);
```

(%o13) $\frac{4\,a^2}{b^2}+a^2$

1.5 USING THE CALCULUS MENU

Calculate Sum: to calculate a summation like:

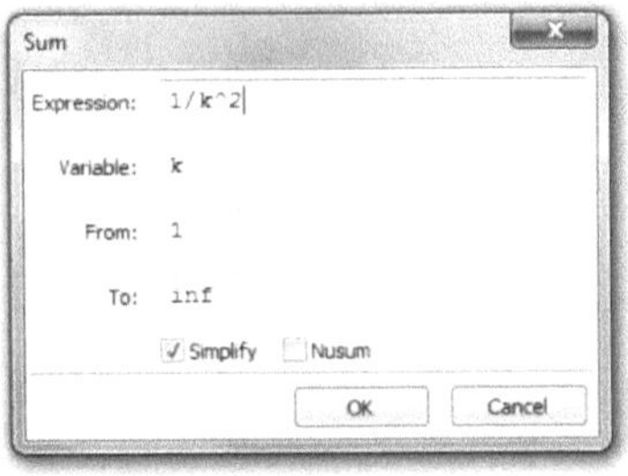


```
(%i7) sum(1/k^2, k, 1, inf), simpsum;
```

(%o7) $\frac{\pi^2}{6}$

Calculate Product: for calculating a product like:

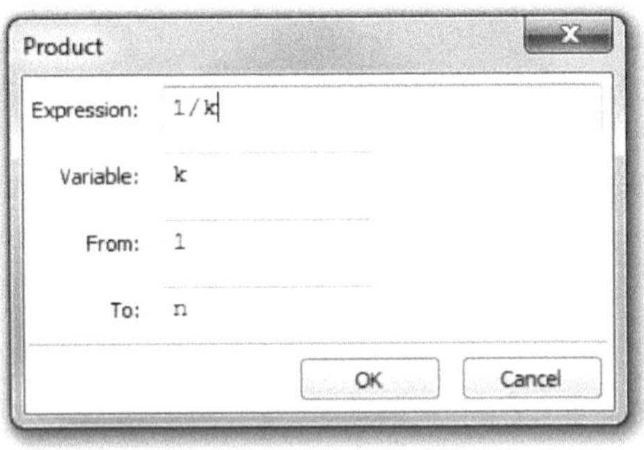


```
(%i8) product(1/k, k, 1, n);
```

(%o8) $\prod_{k=1}^{n} \frac{1}{k}$

Differentiate: calculates a derivative like:

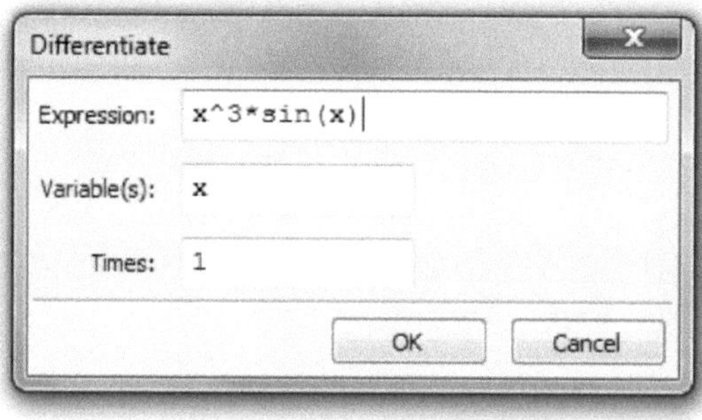


```
(%i9) diff(x^3*sin(x),x,1);
```

(%o9) $3\,x^2\sin(x)+x^3\cos(x)$

Integrate: calculates an integral:

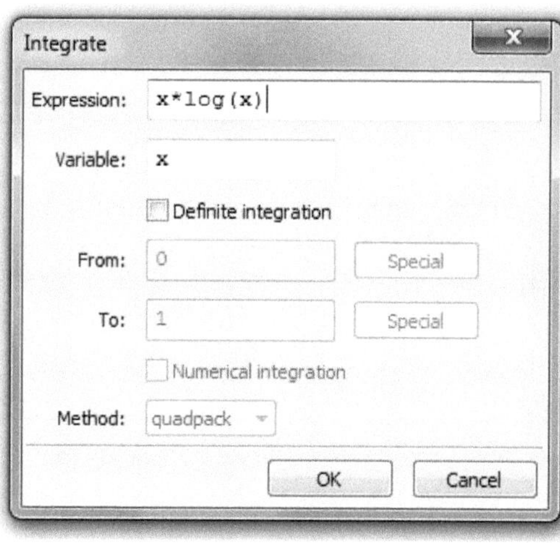


```
(%i10) integrate(x*log(x), x);
```

(%o10) $\frac{x^2\log(x)}{2}-\frac{x^2}{4}$

Find Limit: calculates the limit of a function:

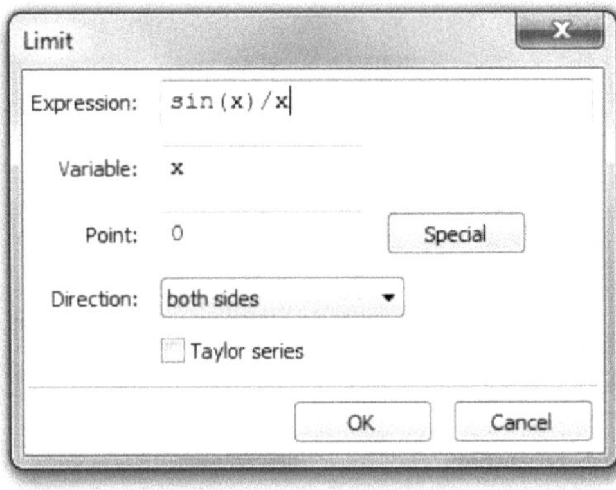


```
(%i11) limit(sin(x)/x, x, 0);
(%o11) 1
```

Get Series: calculates a Taylor series for an expression:

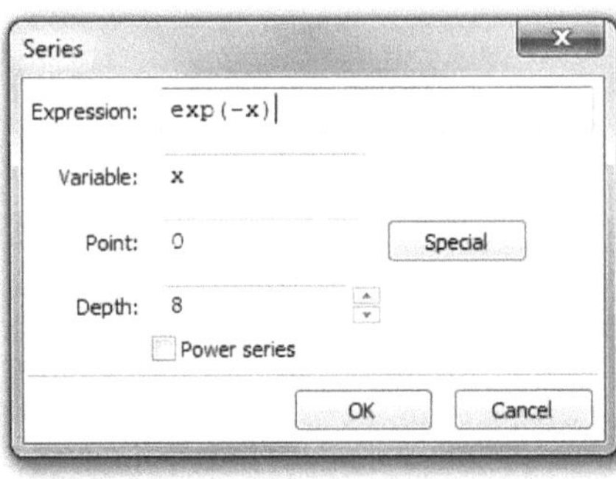


```
(%i12) taylor(exp(-x), x, 0, 8);
```

(%o12)/T/ $1-x+\frac{x^2}{2}-\frac{x^3}{6}+\frac{x^4}{24}-\frac{x^5}{120}+\frac{x^6}{720}-\frac{x^7}{5040}+\frac{x^8}{40320}+...$

1.6 USING THE ALGEBRA MENU

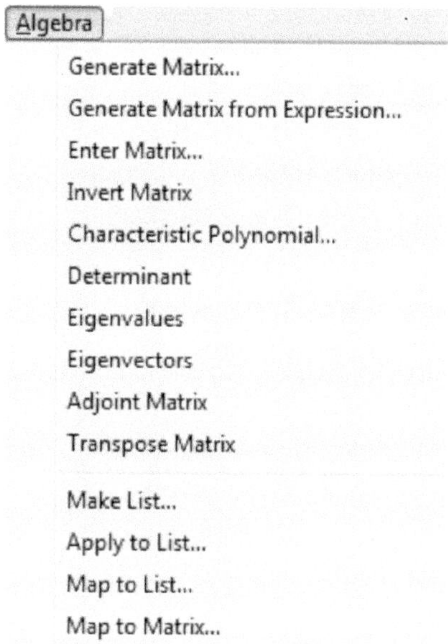

Map to List: maps a function to a list:

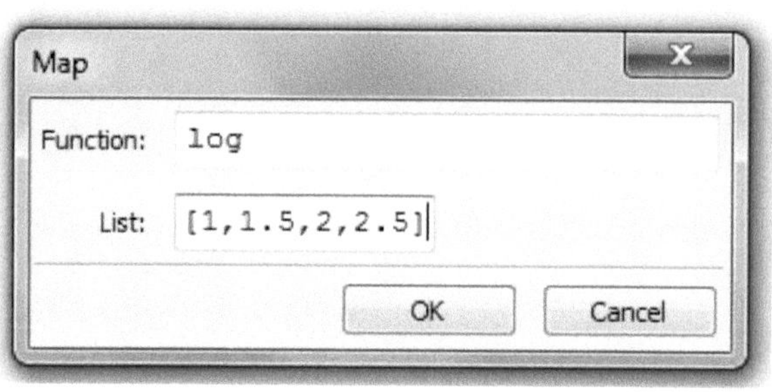


```
(%i14) map(log , [1,1.5,2,2.5]);
(%o14) [0,0.40546510810816,log(2),0.91629073187416]
```

1.7 THE EQUATIONS MENU

A listing of the available applications in the *Equations* menu is shown below:

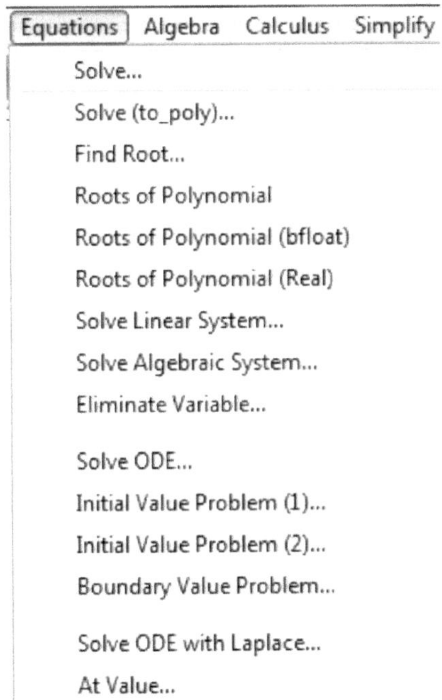

Solve: to solve the equation:

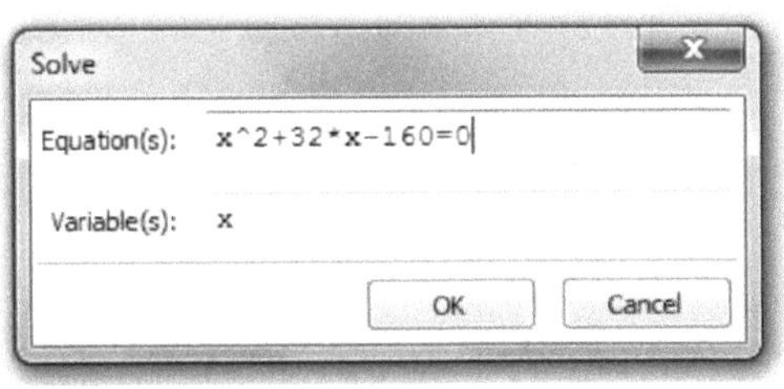


```
(%i1) solve([x^2+32*x-160=0], [x]);
```

(%o1) $[x=-4\sqrt{26}-16, x=4\sqrt{26}-16]$

Find Root: find roots of the equation:

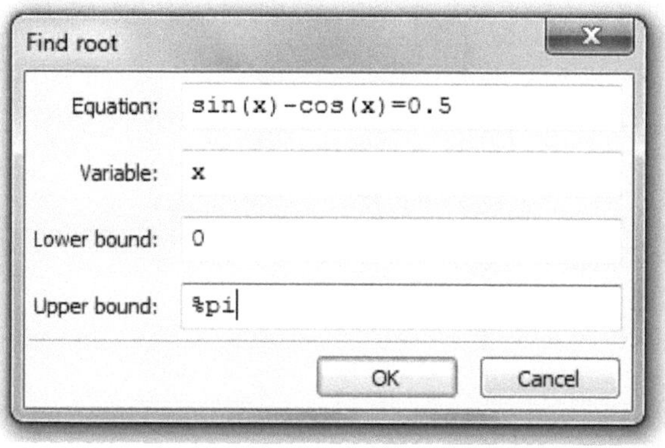


```
(%i5) find_root(sin(x)-cos(x)=0.5, x, 0, %pi);
(%o5) 1.146765287304156
```

Roots of the polynomial: find all roots of a polynomial:

```
-->   x^3+25*x^2-5*x+212=0
```

```
(%i2) realroots(x^3+25*x^2-5*x+212=0);
```

(%o2) $[x=-\frac{856355949}{33554432}]$

Solve linear system: solve linear systems of n equations:

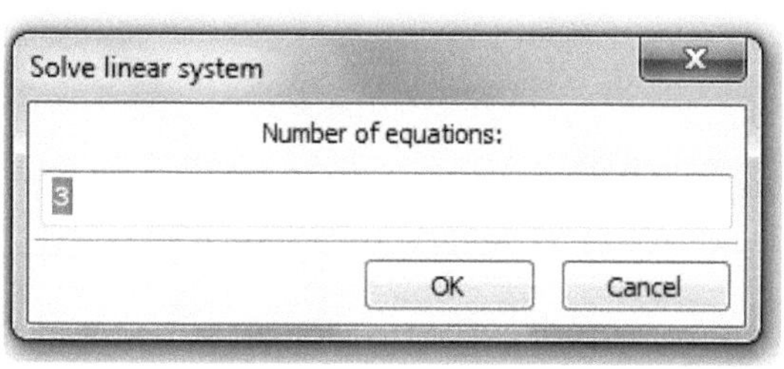

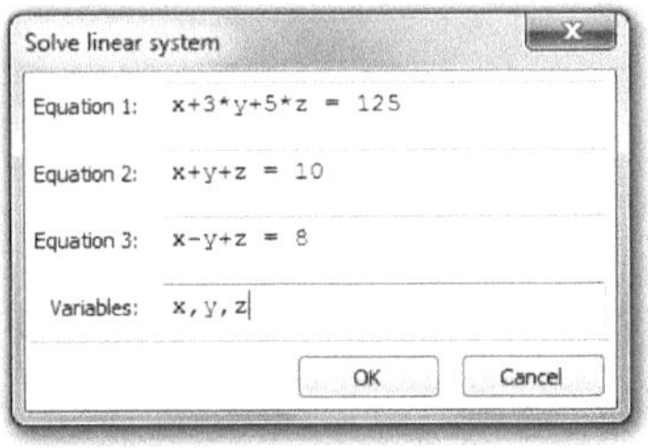


```
(%i3) linsolve([x+3*y+5*z = 125, x+y+z = 10, x-y+z = 8], [x,y,z]);
```

(%o3) $[x=-\frac{77}{4}, y=1, z=\frac{113}{4}]$

Solve algebraic system:

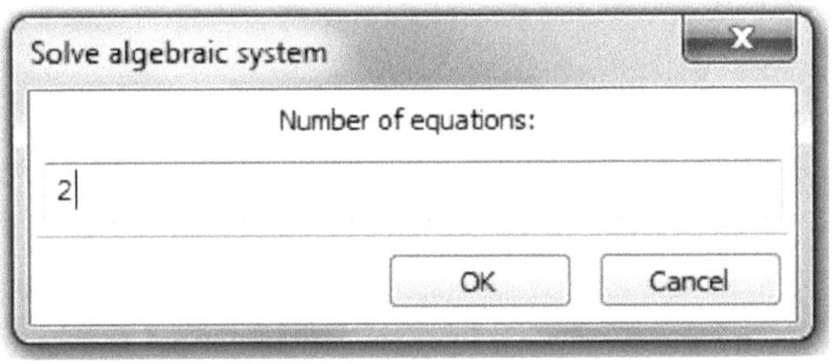

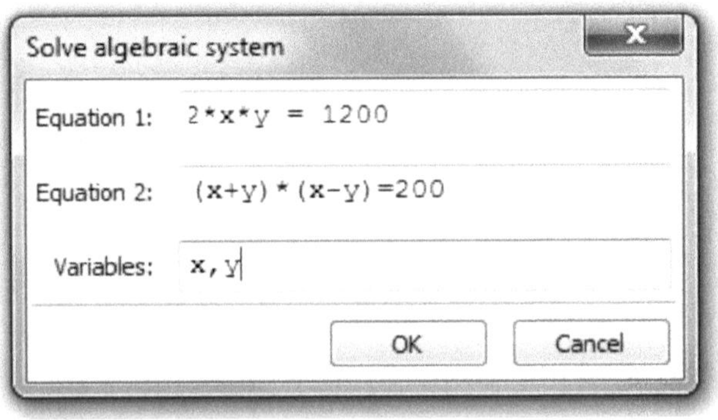


```
(%i4) algsys([x*y^2+2*x*y = 1200 , (x+y)*(x-y)=200], [x,y]);
(%o4) [[x=2.428628665654874 %i-6.665078541235869,y=1.267546365968231-12.77034216552857 %i],[x=-2.428628665654874 %i-
6.665078541235869,y=12.77034216552856 %i+1.26754636596833],[x=2.579774447335989 %i-9.890723428643293,y=10.70113149615758 %i-
2.38440538517289],[x=-2.579774447335989 %i-9.890723428643293,y=-10.70113149615758 %i-2.38440538517289],[x=16.09840425531915,y
=7.691463414634146],[x=17.01320132013201,y=-9.45774647887324]]
```

Solve ODE: solve the ordinary differential equation:

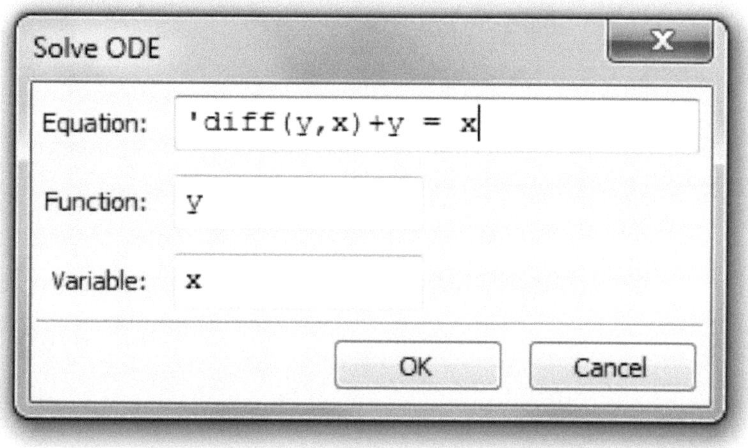


```
(%i5) ode2('diff(y,x)+y=x, y, x);
```

(%o5) $y=\%e^{-x}\left((x-1)\%e^{x}+\%c\right)$

Initial value problem (1): solve initial value problem for the first-order ODE:

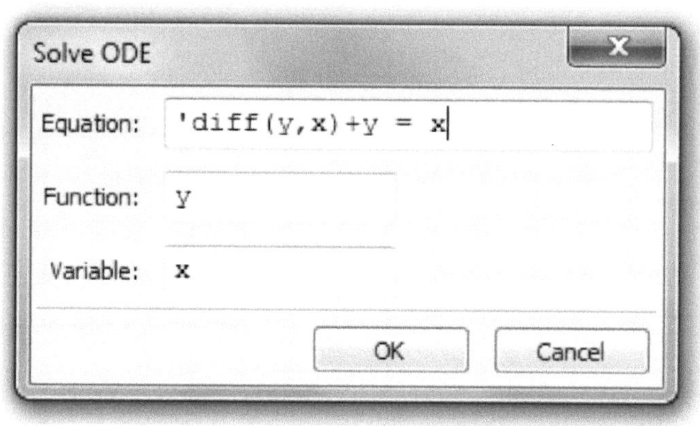

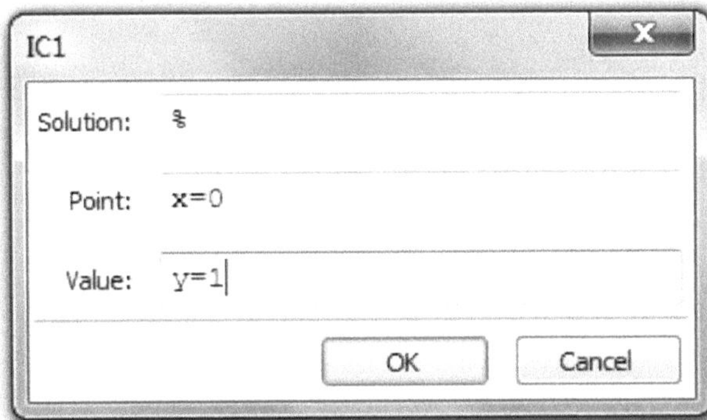


```
(%i1) ode2('diff(y,x)+y =x, y, x);
```

(%o1) $y=\%e^{-x}\left((x-1)\%e^{x}+\%c\right)$

```
(%i2) ic1(%, x=0, y=1);
```

(%o2) $y=\%e^{-x}\left((x-1)\%e^{x}+2\right)$

Initial value problem (2): solve initial value problem for second-order ODE.

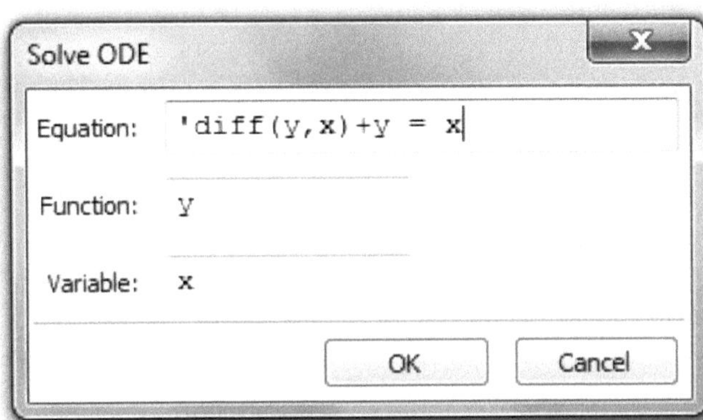

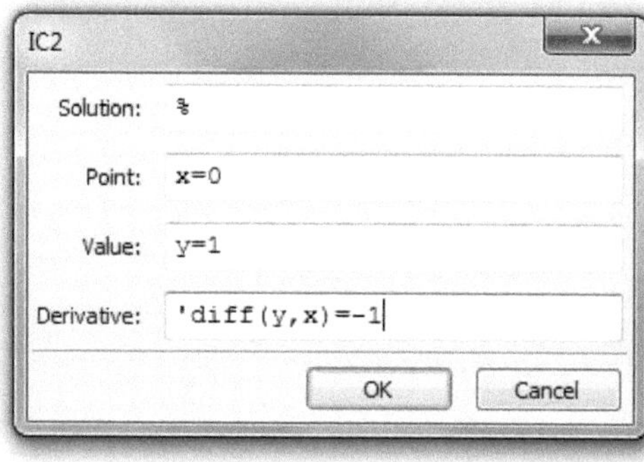


```
(%i3) ode2('diff(y,x)+y =x, y, x);
(%o3) y=%e^-x((x-1)%e^x+%c)

(%i4) ic2(%, x=0, y=1, 'diff(y,x)=-1);
(%o4) []
```

Boundary value problem: solves the boundary value problem for second-order ODE.

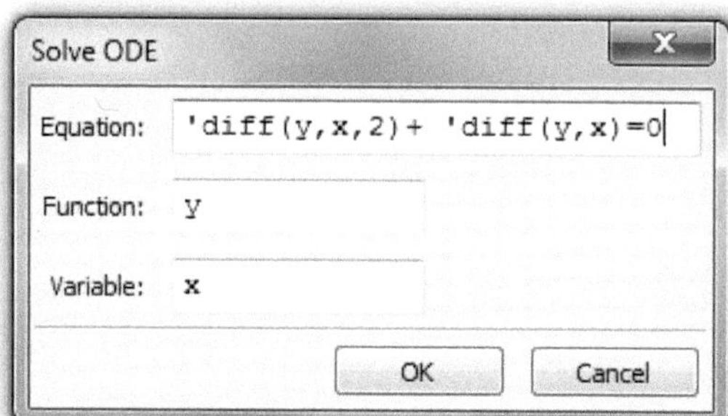

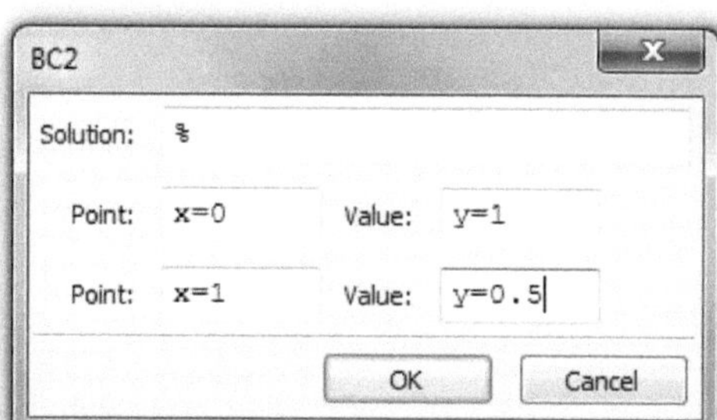

```
(%i5) ode2('diff(y,x,2)+ 'diff(y,x)=0, y, x);
```

(%o5) $y=\%k2\ \%e^{-x}+\%k1$

```
(%i6) bc2(%, x=0, y=1, x=1, y=0.5);
rat: replaced 0.5 by 1/2 = 0.5
```

(%o6) $y=\dfrac{\%e^{1-x}}{2\ \%e-2}+\dfrac{\%e-2}{2\ \%e-2}$

Solve ODE with Laplace: solve an ordinary differential equation using Laplace transforms.

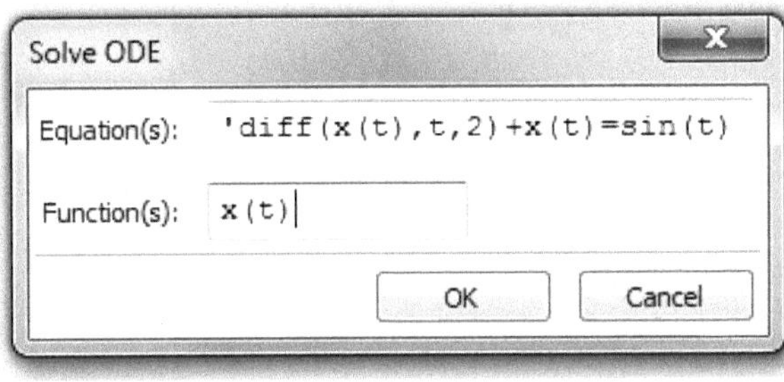


```
(%i7) desolve(['diff(x(t),t,2)+x(t)=sin(t)],[x(t)]);
```

(%o7) $x(t)=\dfrac{\sin(t)\left(2\left(\left.\dfrac{d}{d\,t}x(t)\right|_{t=0}\right)+1\right)}{2}-\dfrac{t\cos(t)}{2}+x(0)\cos(t)$

At value: replace a variable in an expression.

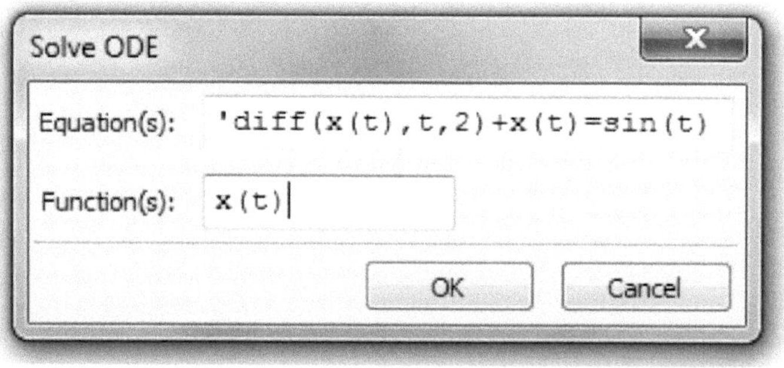

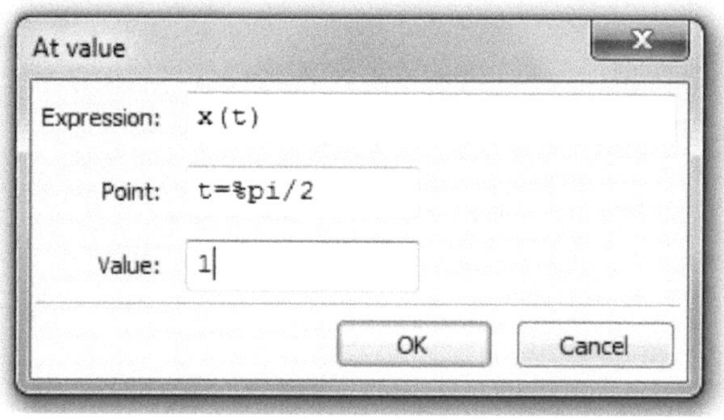


```
(%i7) desolve(['diff(x(t),t,2)+x(t)=sin(t)],[x(t)]);
```

(%o7) $$x(t)=\frac{\sin(t)\left(2\left(\left.\frac{d}{d\,t}x(t)\right|_{t=0}\right)+1\right)}{2}-\frac{t\cos(t)}{2}+x(0)\cos(t)$$

```
(%i8) atvalue(x(t), t=%pi/2, 1);
(%o8) 1
```

1.8 DEFINING FUNCTIONS

To **define a function** we use the below syntax (f(x):=)

```
(%i9) f(x) := x^3+2 $ f(a);f(2);f(1+3/2);
(%o10) a^3+2
(%o11) 10
        141
(%o12) ---
         8
```

Defining of piecewise function: to define piecewise functions in Maxima we can use the *block* statement. Example:

block([<variables with assignments>], <expressions>)

To illustrate the use of the *block* statement in defining a function, consider the function:

$$f(x)=\begin{cases} x+1, \ if \ 0\le x<2 \\ (x+1)^2, \ if \ 2\le x<4 \\ 0, \ elsewhere \end{cases}$$

```
(%i13) f(x):= block(if(0<=x and x<2) then return (x+1), if (2<=x and x<4) then return ((x+1)^2) else return (0));
(%o13) f(x):=block(if 0<=x∧x<2 then return(x+1) ,if 2<=x∧x<4 then return((x+1)^2) else return(0))

(%i14) f(2);
(%o14) 9
```

1.9 COMPLEX NUMBERS

To define and use the complex number in Maxima:

$z=4-\frac{1}{2}i$ and $w=9+\frac{5}{2}i$

```
(%i1) z:4-1/2*%i;
          %i
(%o1) 4 - --
          2

(%i2) w:9+5/2*%i;
       5 %i
(%o2)  ---- + 9
        2
```

To find the conjugate of z and w:

```
(%i9) conjugate(z);
```

(%o9) $\frac{\%i}{2}+4$

```
(%i11) conjugate(w);
```

(%o11) $9-\frac{5\ \%i}{2}$

CHAPTER 2

PRE-CALCULUS

2.1 FUNDAMENTAL OPERATIONS WITH NUMBERS

Perform each of the indicated operations using Maxima

1. 42 + 23,23 + 42

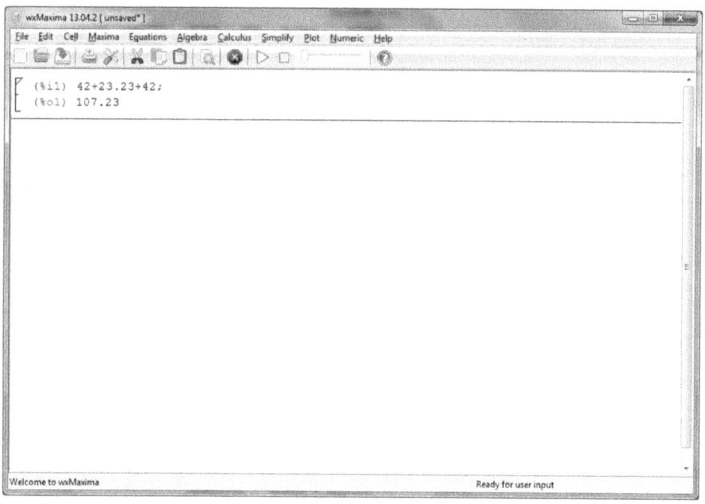

NB: In Maxima, for the decimal numbers you should write them in this way 23.23 for 23,23

You have to write the expression in Maxima and press **Shift Enter** to execute the demand.

```
(%i2) 42 + 23.23 + 42;
(%o2) 107.23
```

2. 27 + (48 + 12)

```
(%i2) 42 + 23.23 + 42;
(%o2) 107.23
```

3. (27 + 48) + 12

```
(%i4) (27 + 48) + 12;
(%o4) 87
```

4. 125 – (38 + 27)

```
(%i5) 125 -(38+ 27);
(%o5) 60
```

5. 6 × 8

NB: In Maxima the multiplication sign is replaced by *

```
(%i6) 6 * 8;
(%o6) 48
```

6. 8 × 6

```
(%i7) 8 * 6;
(%o7) 48
```

7. 4(7 × 6)

```
(%i8) 4 * (7 * 6);
(%o8) 168
```

8. (4 × 7)6

```
(%i9) (4*7)*6;
(%o9) 168
```

9. $35 - 28$

```
(%i10) 35- 28;
(%o10) 7
```

10. $756 \div 21$

NB: In Maxima the multiplication sign is replaced by /

```
(%i12) 756/21;
(%o12) 36
```

11. $\frac{(40+21)(72-38)}{(32-15)}$

```
(%i13) ((40+21)*(72-38))/(32-15);
(%o13) 122
```

12. $72 \div 24 + 64 \div 16$

```
(%i14)  72/24+64/16;
(%o14) 7
```

13. $4 \div 2 + 6 \ \div 3 - 2 \div 2 + 3.4$

```
(%i15) 4/2+6/3-2/2+3.4;
(%o15) 6.4
```

14. $128 \div (2 \times 4)$

```
(%i16) 128/(2*4);
(%o16) 16
```

15. $(128 \div 2) \times 4$

```
(%i17) (128/2)*4;
(%o17) 256
```

16. $(5)(-3)(-2)$

```
(%i18) 5 * (-3)*(-2);
(%o18) 30
```

17. $\dfrac{8(-2)}{-4}+\dfrac{(-4)(-2)}{2}$

```
(%i19) (8*(-2))/(-4)+((-4)*(-2))/2;
(%o19) 8
```

18. $\dfrac{12(-40)(-12)}{5(-3)-3(-3)}$

```
(%i20) (12 *(-40)*(-12))/(5*(-3)-3*(-3));
(%o20) -960
```

19. 2^3

NB: In Maxima, you should use this sign ^ for the exponent.

```
(%i21) 2^3;
(%o21) 8
```

20. $5(3^2)$

```
(%i22) 5 * (3^2);
(%o22) 45
```

21. $2^6 \times 2^4$

```
(%i23) 2^6 * 2^4;
(%o23) 1024
```

22. $\frac{3^4 \times 3^3}{3^2}$

```
(%i24) (3^4 *3^3)/3^2;
(%o24) 243
```

APPLICATION

Perform each of the following operations using Maxima

1) $40 + (30 \times 2)$
2) $4^5 + 5^2$
3) $80 \div 2 + 30 \div 10$
4) $125 - (50 + 25)$
5) $\frac{(30+2)(10\times5)}{50\times2}$
6) $3(-2)(-4)$
7) $4(3)^3$
8) $\frac{3(-4)}{(-2)(-6)}$
9) $32 - (9 \times 4)$
10) $(10 + 2) - (5 \times 2)$

2.2 SOME PRODUCTS

Find each of the following products

1. $a(c + d)$

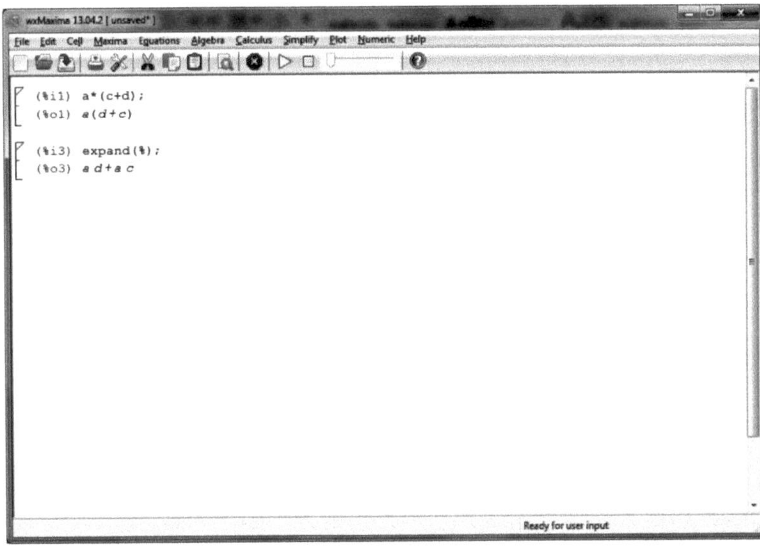

Steps:

1) Go to Maxima;
2) Enter your command: a ∗ $(c + d)$;
3) Go to the menu bar click on Simplify → Expand Expression;
4) Press Shift Enter to execute your command.

Or you can type expand (a ∗ $(c + d)$) then press Shift Enter

```
--> a * (c +d)

(%i2) expand(a * (c +d));
(%o2) a d+a c
```

2. $(a + b)\,(a - b)$

```
-->   (a+b)*(a-b)

(%i3) expand((a+b)*(a-b));
```
(%o3) a^2-b^2

3. $(a+b)^2$

```
-->   (a+b)^2

(%i4) expand((a+b)^2);
```
(%o4) $b^2+2\,a\,b+a^2$

4. $(a-b)^2$

```
-->   (a-b)^2

(%i6) expand((a-b)^2);
```
(%o6) $b^2-2\,a\,b+a^2$

5. $(x+a)(x+b)$

```
-->   (x+a)*(x+b)

(%i7) expand((x+a)*(x+b));
```
(%o7) $x^2+b\,x+a\,x+a\,b$

6. $(a+b)^3$

```
(%i9) expand ((a+b)^3);
```
(%o9) $b^3+3\,a\,b^2+3\,a^2\,b+a^3$

7. $(a - b)^3$

```
(%i10) expand ((a-b)^3);
```

(%o10) $-b^3+3\,a\,b^2-3\,a^2\,b+a^3$

8. $(a + b + c)^2$

```
(%i11) expand ((a+b+c)^2);
```

(%o11) $c^2+2\,b\,c+2\,a\,c+b^2+2\,a\,b+a^2$

9. $(a - b)(a^2 + ab + b^2)$

```
-->    (a-b)  *  (a^2 + ab + b^2)

(%i12) expand((a-b)  *  (a^2 + ab + b^2));
```

(%o12) $-b^3+a\,b^2-ab\,b-a^2\,b+a\,ab+a^3$

10. $3x(2x + 3y)$

```
-->    3*x  *  (2*x + 3 *y)

(%i14) expand(3*x  *  (2*x + 3 *y));
```

(%o14) $9\,x\,y+6\,x^2$

11. $x^2y(3x^3 - 2y + 4)$

```
(%i15) expand (x^2*y (3*x^3 - 2*y +4));
```

(%o15) $x^2\,y(-2\,y+3\,x^3+4)$

12. $(2x + 3y)(2x - 3)$

```
(%i18) expand ((2*x-3*y)*(2*x + 3*y));
(%o18) 4 x^2 - 9 y^2
```

13. $(1 - 5x^3)(1 + 5x^3)$

```
(%i19) expand ((1-5*x^3)*(1+5*x^3));
(%o19) 1-25 x^6
```

14. $(3x + 5y)^2$

```
(%i20) expand ((3*x+5*y)^2);
(%o20) 25 y^2 +30 x y+9 x^2
```

15. $(ax - 2by)^2$

```
(%i21) expand ((a*x-2*b*y)^2);
(%o21) 4 b^2 y^2 -4 a b x y+a^2 x^2
```

16. $(x + 3)(x + 5)$

```
(%i24) expand ((x+3)*(x+5));
(%o24) x^2 +8 x+15
```

17. $(x - 2)(x + 8)$

```
(%i25) expand ((x-2)*(x+8));
(%o25) x^2 +6 x -16
```

APPLICATION

Find each of the following products

1) $3x(5 + 2x)$
2) $4(10x - 3)$
3) $(2x + 3)^2$
4) $(2x - 3)^2$
5) $(x - 3)(x + 4)$
6) $(x - 3)^3$
7) $(x + 4)^3$
8) $(2x + 3 + y)^2$
9) $(2x - 3y)(x - 4)$
10) $(x - 4)(3x + 2)$

2.3 FACTORING

Factorize

1. $ac + ad$

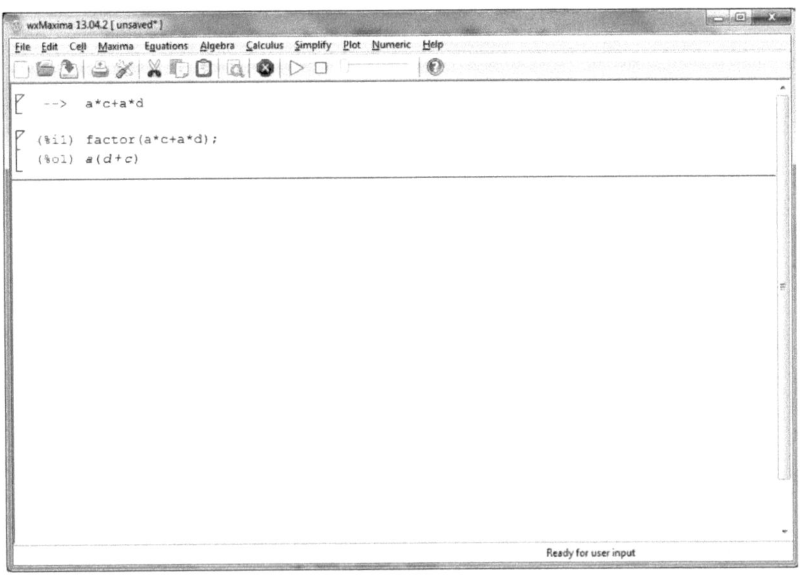

Steps:

1) Go to Maxima
2) Enter your command: $a* c + a * d$
3) Go to the menu bar click on Simplify → Factor Expression

4) Press Shift Enter to execute your command
Or you can type factor ($a*c + a*d$) then press Shift Enter

```
-->   a*c + a *d

(%i1) factor(a*c + a *d);
(%o1) a(d+c)
```

2. $2x^2 - 3xy$

```
(%i2) factor(2*(x^2) - 3*x*y);
(%o2) -x(3 y-2 x)
```

3. $4x + 8y + 12z$

```
(%i3) factor(4*x+8*y+12*z);
(%o3) 4(3 z+2 y+x)
```

4. $9s^3t + 15s^2t^3 - 3s^2r^2$

```
-->   9*s^3*t + 15*s^2*t^3 -3*s^2*r^2

(%i4) factor(9*s^3*t + 15*s^2*t^3 -3*s^2*r^2);
(%o4) 3 s^2 (5 t^3+3 s t-r^2)
```

5. $4a^{n+1} - 8a^{2n}$

```
-->   4*a^(n+1)-8*a^(2*n)

(%i5) factor(4*a^(n+1)-8*a^(2*n));
(%o5) -4 a^n (2 a^n-a)
```

6. $a^2 - b^2$

```
(%i6) factor(a^2 - b^2);
(%o6) -(b-a)(b+a)
```

7. $x^2 - 9$

```
(%i7) factor(x^2 - 9);
(%o7) (x-3)(x+3)
```

8. $25x^2 - 4y^2$

```
-->   25 * x^2 - 4 * y^2

(%i8) factor(25 * x^2 - 4 * y^2);
(%o8) -(2 y-5 x)(2 y+5 x)
```

9. $x^4 - y^4$

```
(%i10) factor(x^4 - y^4);
(%o10) -(y-x)(y+x)(y^2+x^2)
```

10. $(x + 1)^2 - 36y^2$

```
-->   (x+1)^2 - 36*y^2

(%i11) gfactor((x+1)^2 - 36*y^2);
(%o11) -(6 y-x-1)(6 y+x+1)
```

11. $a^2 + 2ab + b^2$

```
(%i13) factor(a^2 + 2*a*b + b^2);
```

$$(\%o13)\ (b+a)^2$$

12. $a^2 - 2ab + b^2$

```
(%i14) factor(a^2 - 2*a*b + b^2);
```

$$(\%o14)\ (b-a)^2$$

13. $x^2 + 8x + 16$

```
(%i15) factor(x^2 + 8*x + 16);
```

$$(\%o15)\ (x+4)^2$$

14. $1 + 4y + 4y^2$

```
(%i16) factor(1+4*y+4*y^2);
```

$$(\%o16)\ (2\,y+1)^2$$

15. $x^2 - 4x + 4$

```
(%i17) factor(x^2 - 4*x + 4);
```

$$(\%o17)\ (x-2)^2$$

16. $16m^2 - 40mn + 25n^2$

```
--> 16*m^2-40*m*n+25*n^2

(%i18) factor(16*m^2-40*m*n+25*n^2);
```

$$(\%o18)\ (5\,n-4\,m)^2$$

17. $x^2 + 6x + 8$

```
 -->    x^2+6*x+8

(%i21) factor(x^2+6*x+8);
(%o21) (x+2)(x+4)
```

18. $x^2 - 6x + 8$

```
(%i22) factor(x^2-6*x+8);
(%o22) (x-4)(x-2)
```

19. $z^4 - 10z^2 + 9$

```
 -->    z^4-10*z^2+9

(%i23) factor(z^4-10*z^2+9);
(%o23) (z-3)(z-1)(z+1)(z+3)
```

20. $3x^3 - 3x^2 - 18x$

```
 -->    3*x^3-3*x^2-18*x

(%i24) factor(3*x^3-3*x^2-18*x);
(%o24) 3(x-3)x(x+2)
```

21. $(x + 1)^2 + 3(x + 1) + 2$

```
(%i25) factor((x+1)^2+3*(x+1)+2);
(%o25) (x+2)(x+3)
```

22. $x^{2a} - x^a - 30$

```
(%i26) factor(x^(2*a)-x^a -30);
```

(%o26) $(x^a - 6)(x^a + 5)$

23. $3x^2 + 10x + 3$

```
(%i27) factor(3*x^2+10*x+3);
```

(%o27) $(x+3)(3\,x+1)$

24. $a^3 + b^3$

```
(%i28) factor(a^3+b^3);
```

(%o28) $(b+a)(b^2 - a\,b + a^2)$

25. $a^3 - b^3$

```
(%i29) factor(a^3-b^3);
```

(%o29) $-(b-a)(b^2 + a\,b + a^2)$

26. $a^6 - b^6$

```
(%i30) factor(a^6-b^6);
```

(%o30) $-(b-a)(b+a)(b^2 - a\,b + a^2)(b^2 + a\,b + a^2)$

27. $64x^3 + 125y^3$

```
(%i31) factor(64*x^3+125*y^3);
```

(%o31) $(5\,y + 4\,x)(25\,y^2 - 20\,x\,y + 16\,x^2)$

28. $ac + bc + ad + bd$

```
(%i32) factor(a*c+b*c+a*d+b*d);
(%o32) (b+a)(d+c)
```

29. $bx - ab + x^2 - ax$

```
(%i33) factor(b*x-a*b+x^2-a*x);
(%o33) (x-a)(x+b)
```

30. $a^6 + b^6 - a^2b^4 - a^4b^2$

```
(%i34) factor(a^6+b^6-a^2*b^4-a^4*b^2);
(%o34) (b-a)^2 (b+a)^2 (b^2+a^2)
```

APPLICATION

Factorize using Maxima

1) $2x + 4x^2$
2) $3r^2 + 9r^2 + 6rt$
3) $9x^2 - 25$
4) $x^2 - 4$
5) $4x^2 + 32x + 36$
6) $4x^2 - 32x + 36$
7) $1 + 4x + 4x^2$
8) $2x^2 - 12x + 16$
9) $3x^2 + 11x + 6$
10) $x^3 - x^2 - 3x$

2.4 OPERATIONS WITH POLYNOMIALS

1. Give the absolute values of the following using Maxima:

NB: To find the absolute value of a number you need to use this command “abs”

a) 7

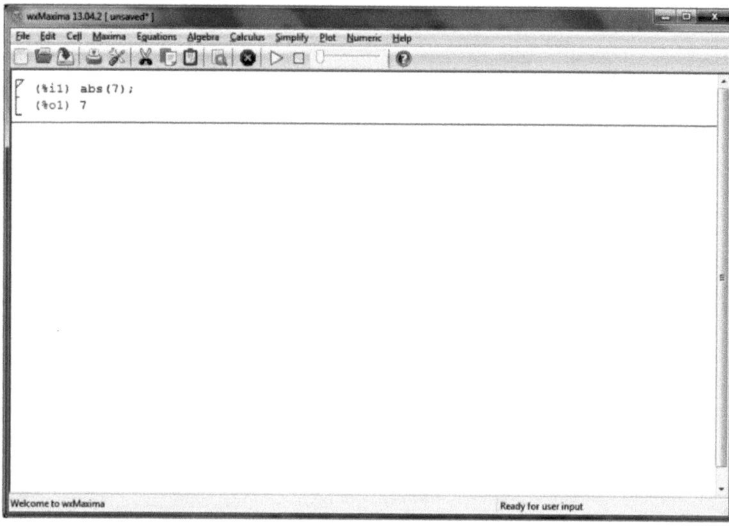


```
(%i1) abs(7);
(%o1) 7
```

b) –15

```
(%i2) abs(-15);
(%o2) 15
```

c) +3 ½

```
(%i3) abs(+3.5);
(%o3) 3.5
```

d) –v

```
(%i4) abs(-v);
(%o4) |v|
```

2. Perform each of the indicated operations and simplify using Maxima
   a) $(3y)(-y)$

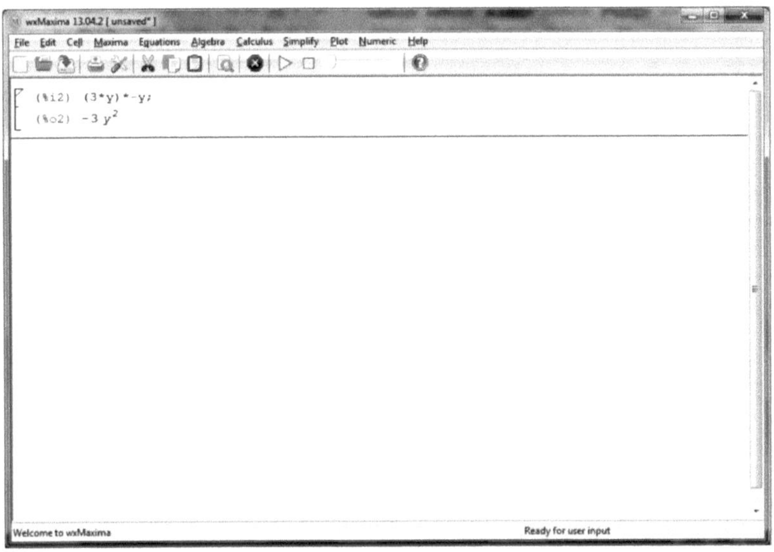


```
(%i5) 3*y * -y;
(%o5) -3 y^2
```

b) $(4a)^2$

```
(%i6) (4*a)^2;
(%o6) 16 a^2
```

c) $-32t \div 4$

```
(%i7) -32*t/4;
(%o7) -8 t
```

d) $3t - 5\,[(2t + 1) - (4 - t)]$

```
(%i8) 3*t -5*((2*t+1)-(4-t));
(%o8) 3 t-5(3 t-3)
```

e) $(2x^2 - 3x + 1) + (5x^2 + 7x - 4)$

```
(%i9) ((2*x^2-3*x+1)+(5*x^2+7*x-4));
(%o9) 7 x^2+4 x-3
```

f) $(2x^2 - 4xy + 4y^2) \div 2$

```
(%i10) ((2*x^2-4*x*y+4*y^2)/2);
```

(%o10) $\frac{4\,y^2-4\,x\,y+2\,x^2}{2}$

Click on →Simplify expression

```
(%i11) ratsimp(%);
(%o11) 2 y^2-2 x y+x^2
```

g) $(2x - 3y)\,(5x + y)$

```
(%i12) ((2*x -3*y)*(5*x+y));
(%o12) (2 x-3 y)(y+5 x)
```

```
(%i13) expand(%);
(%o13) -3 y^2-13 x y+10 x^2
```

h) $\frac{3}{x-1} + \frac{2}{1-x}$

```
(%i14) ((3/(x-1))+(2/(1-x)));
```

$$(\%o14)\ \frac{3}{x-1}+\frac{2}{1-x}$$

```
(%i15) ratsimp(%);
```

$$(\%o15)\ \frac{1}{x-1}$$

i) $\dfrac{3ab}{2c} \div \dfrac{3c}{2ab}$

```
(%i16) (((3*a*b)/2*c)/(3*c)/(2*a*b));
```

$$(\%o16)\ \frac{1}{4}$$

Application

1) Give the absolute values of the following using Maxima
   a) -3
   b) 24
   c) $3-6$
   d) $-y-z$
   e) $32-20$
2) Perform each of the following operations and simplify using Maxima
   a) $-y(5xy)$
   b) $(6b)^2$
   c) $-24s \div 12$
   d) $(4x^2+5x+3)-(3x^2+6x-3)$
   e) $(3x+2)(3y+1)$

2.5 LINEAR EQUATIONS

Solve the following equations using Maxima:

1. $3h = 12$

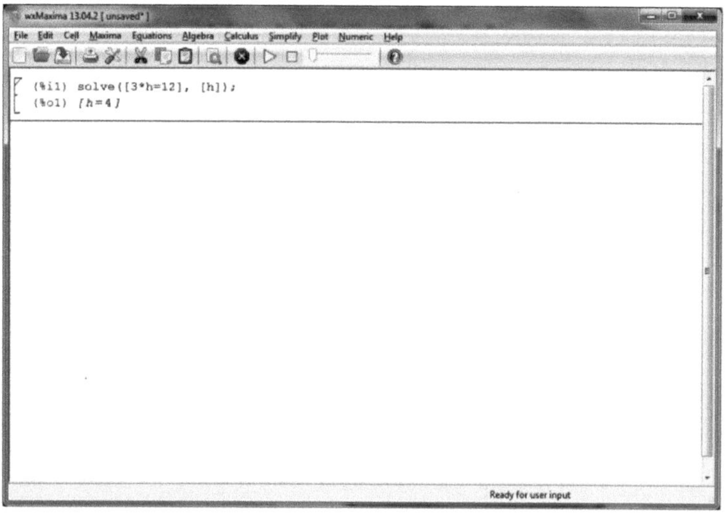

In Maxima, you have to click on Equations → Solve

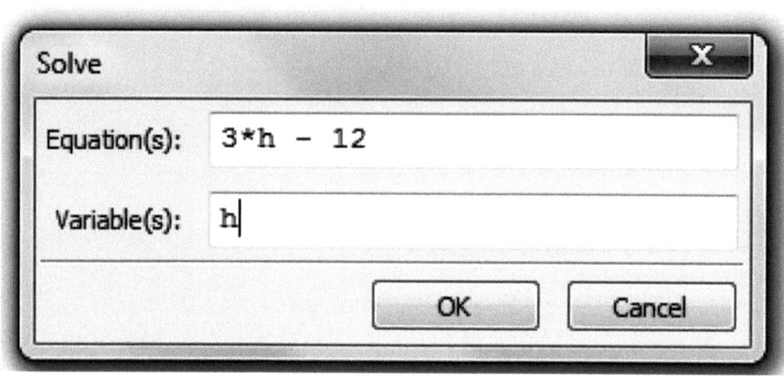


```
(%i1) solve([3*h - 12], [h]);
(%o1) [h=4]
```

Or use this command "Solve"

2. $40 = 8a$

```
(%i3) solve([8*a-40], [a]);
(%o3) [a=5]
```

3. $\frac{y}{3} = 12$

```
(%i4) solve([y/3 -12], [y]);
(%o4) [y=36]
```

4. $a + 6 = 8$

```
(%i6) solve([a+6-8], [a]);
(%o6) [a=2]
```

5. $16 = 2(t + 3)$

```
(%i8) solve([2*(t+3)-16], [t]);
(%o8) [t=5]
```

APPLICATION

Solve the following equations using Maxima

1) $4x = 12$
2) $\frac{z}{2} = 6$
3) $3 + b = 9$
4) $x + 3 = 7$
5) $2x + 10 = 12$
6) $8 + 2a = 5$
7) $3x - 9 = 0$

8) $10 + 2c = 0$
9) $16 = 2(x + 6)$
10) $3a - 2 = 10$

2.6 LINEAR INEQUALITIES

Inequality operators in maxima are as follows:

Symbol	**Description**
<	less than
<=	less than or equal to
>	greater than
>=	greater than or equal to
#	is not

Solve each inequality in Maxima

1. $a - 12 < 6$

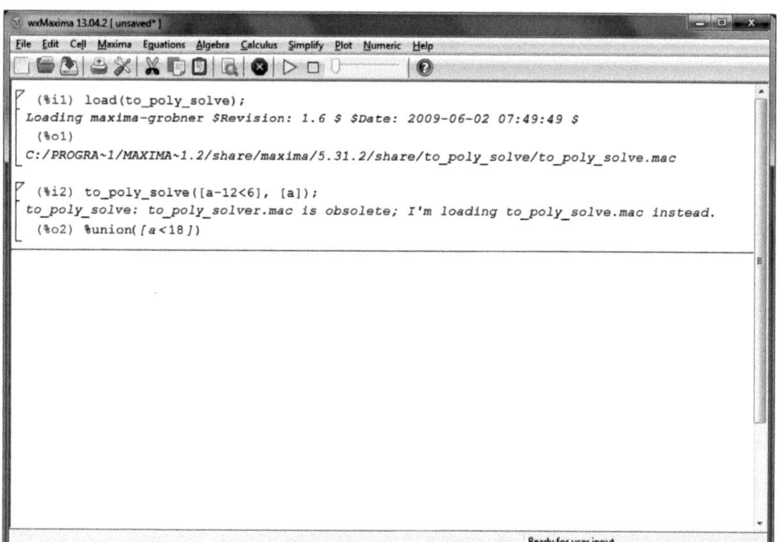

Steps:

1) Go to Maxima
2) Type **load(to_poly_solve);**
3) Click on equations → Solve (to_poly)
4) Type your command

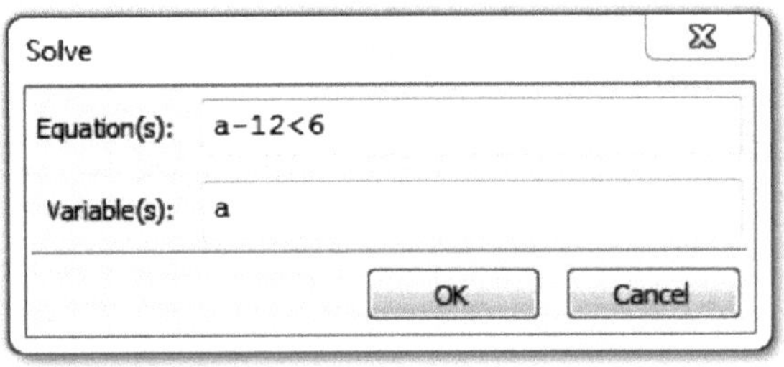

5. Click OK

```
(%i1) load(to_poly_solve);
Loading maxima-grobner $Revision: 1.6 $ $Date: 2009-06-02 07:49:49 $
(%o1)
C:/PROGRA~1/MAXIMA~1.2/share/maxima/5.31.2/share/to_poly_solve/to_poly_solve.mac

(%i2) to_poly_solve([a-12<6], [a]);
to_poly_solve: to_poly_solver.mac is obsolete; I'm loading to_poly_solve.mac instead.
(%o2) %union([a<18])
```

So here the solution is $a < 18$

2. $2x \leq x + 1$

```
(%i3) to_poly_solve([2*x<x+1], [x]);
(%o3) %union([x<1])
```

3. $x + \frac{1}{3} > 4$

```
(%i4) to_poly_solve([x+1/3>4], [x]);
```

(%o4) $\%union\left([\frac{11}{3} < x]\right)$

4. $2x + 3 > x + 5$

```
(%i2) to_poly_solve([2*x-x<=1], [x]);
to_poly_solve: to_poly_solver.mac is obsolete; I'm loading to_poly_solve.mac instead.
(%o2) %union([x=1],[x<1])
```

5. $x+\frac{1}{8}<\frac{1}{2}$

```
(%i6) to_poly_solve([x+1/8<1/2], [x]);
```

(%o6) $\%union\left(\left[x<\frac{3}{8}\right]\right)$

6. $3x-9<2x+6$

```
(%i7) to_poly_solve([3*x-9<=2*x+6], [x]);
(%o7) %union([x=15],[x<15])
```

7. $-0.17x-0.23<0.75-1.17x$

```
(%i9) to_poly_solve([(-0.17)*x-(0.23)<(0.75)-(1.17)*x], [x]);
```

(%o9) $\%union([x<1.1102230246251565\ 10^{-16}\ x+0.98])$

8. $3(r-2)<2r+4$

```
(%i12) to_poly_solve([3*(r-2)<2*r+4], [r]);
(%o12) %union([r<10])
```

APPLICATION

Solve each of the following inequalities using Maxima

1) $x-4\leq 3$
2) $x+3>5$
3) $2x-6\geq 3$
4) $x+\frac{1}{4}\leq\frac{1}{2}$

5) $2x + 4 \geq x + 3$
6) $x + 7 > 2$
7) $2x - 8 > 10$
8) $3x + 6 > 0$
9) $2x + 6 \leq 3$
10) $3x + 6 \leq 2x + 6$

2.7 QUADRATIC EQUATIONS

Solve using Maxima

1. $x^2 - 6 = 0$

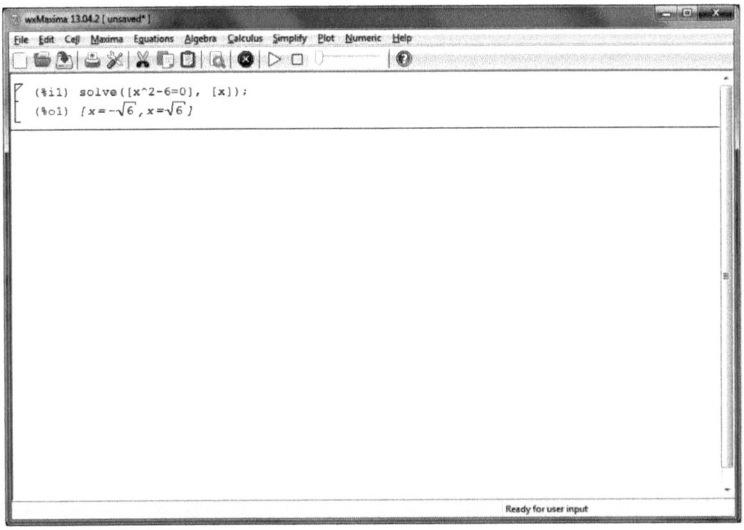


```
(%i1) solve([x^2-6], [x]);
```
(%o1) $[x=-\sqrt{6}, x=\sqrt{6}]$

2. $4t^2 - 9 = 0$

```
(%i2) solve([4*t^2-9], [t]);
```
(%o2) $[t=-\frac{3}{2}, t=\frac{3}{2}]$

3. $4x^2 + 9 = 0$

```
(%i3) solve([4*x^2+9], [x]);
              3 %i     3 %i
(%o3) [x = - ----, x = ----]
               2        2
```

4. $x^2 + 5x - 6 = 0$

```
(%i4) solve([x^2+5*x-6], [x]);
(%o4) [x=-6,x=1]
```

5. $t^2 = 4t$

```
(%i2) solve([t^2=4*t], [t]);
(%o2) [t=0,t=4]
```

6. $\frac{1}{t-1}+\frac{1}{t-4}=\frac{5}{4}$

```
-->  ((1/(t-1))+(1/(t-4))-5/4)
```

```
(%i6) solve([((1/(t-1))+(1/(t-4))-5/4)], [t]);
          8
(%o6) [t=-,t=5]
          5
```

APPLICATION

Solve each of the following equations using Maxima

1) $x^2 - 6x + 8 = 0$
2) $x^2 - 9 = 0$
3) $2x - 4x^2 = 1$
4) $\sqrt{2x+1} = 3$

5) $t^2 = 4 - 3t$
6) $x^2 + 1 = 0$
7) $4t^2 - 25 = 0$
8) $x^2 + 2x + 1 = 0$
9) $x^2 - 2x - 3 = 0$
10) $10x^2 + 5x = 0$

2.8 QUADRATIC FUNCTIONS

Graph each function using Maxima

1. $f(x) = x^2 - 4$

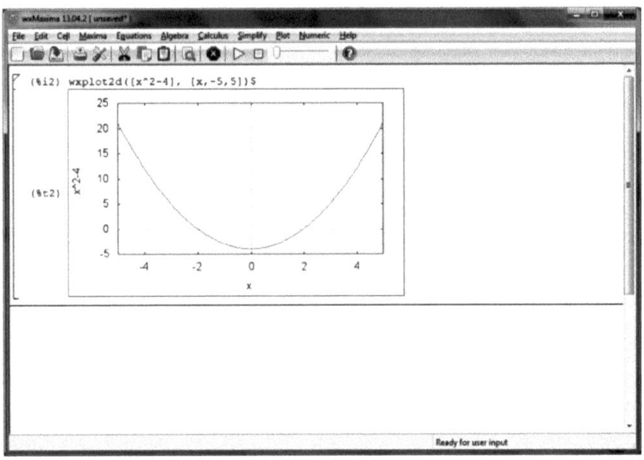

Steps:

1) Go to Maxima
2) Click on plot → plot 2D
3) Type the expression

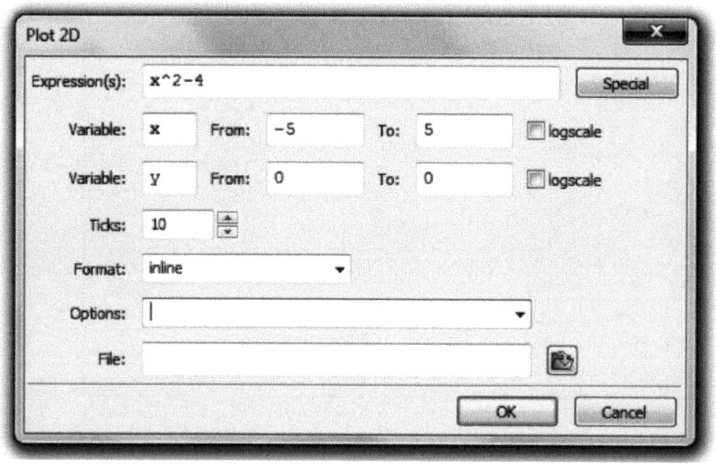

4. Click OK

```
(%i1) wxplot2d([x^2-4], [x,-5,5])$
```

(%t1)

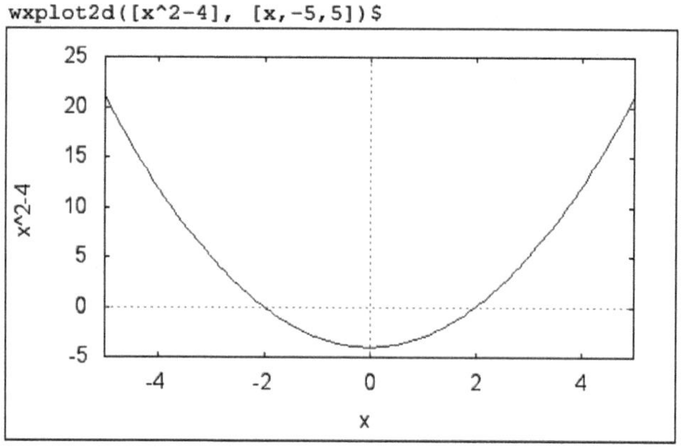

2. $h(x) = -2x^2$

```
(%i2) wxplot2d([-2*x^2], [x,-5,5])$
```

(%t2)

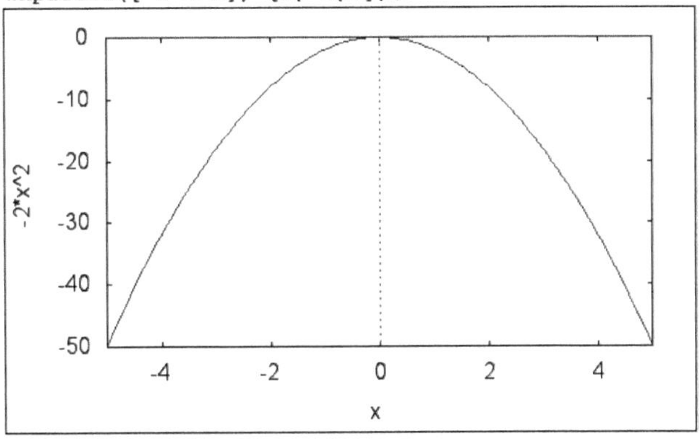

3. $k(x) = x^2 + 2x + 1$

```
(%i3) wxplot2d([x^2+2*x+1], [x,-5,5])$
```

(%t3)

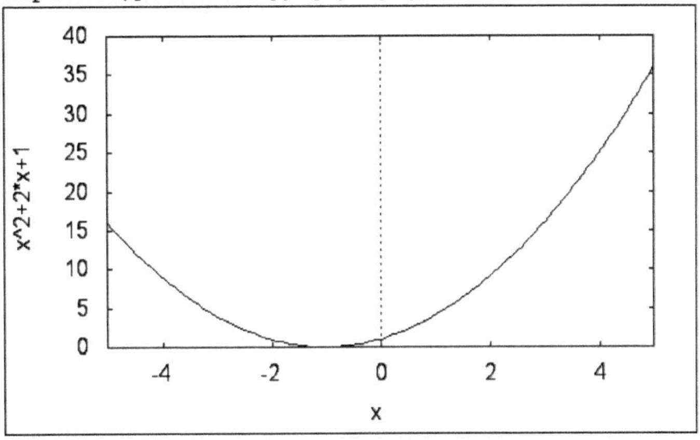

4. $f(x) = x^2 - 2x - 6$

```
(%i4) wxplot2d([x^2-x-6], [x,-5,5])$
```

(%t4)

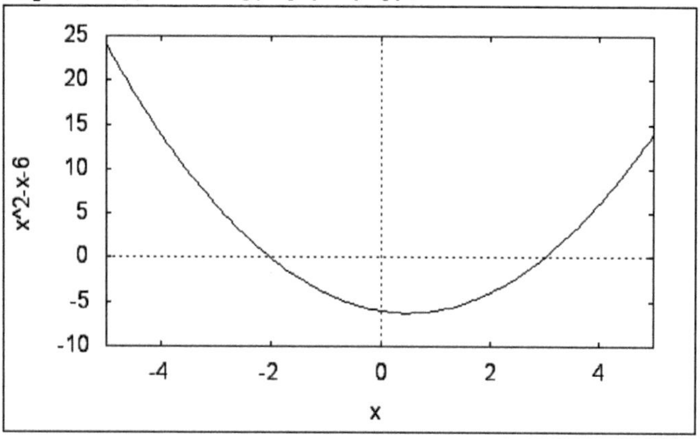

5. $h(x) = 2x^2 + 4x + 1$

```
(%i5) wxplot2d([2*x^2+4*x+1], [x,-5,5])$
```

(%t5)

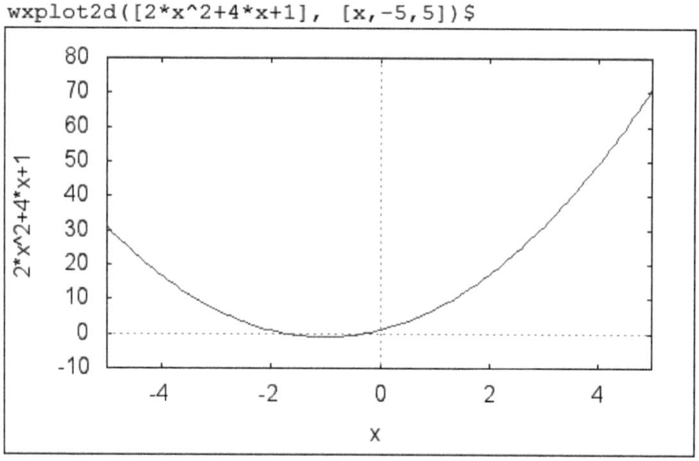

APPLICATION

Graph each of the following functions using Maxima

1) $f(x) = x^2 - 9$
2) $f(x) = x^2 + 4x + 2$
3) $f(x) = x^2 + 4x + 2$
4) $f(x) = -2x^2 - 8x - 4$

5) $f(x) = 2 - x - \frac{1}{2}x^2$

6) $f(x) = 3x^2 + 7x + 1$

7) $y = \frac{2}{3}x^2 - 11$

8) $f(x) = 2x^2 - 4x + 1$

9) $f(x) = \sqrt{x}$

10) $g(x) = x^2 + 3$

2.9 SYSTEM OF LINEAR EQUATIONS

Solve the following systems using Maxima

1. $\begin{cases} 2x - y = 4 \\ x + y = 5 \end{cases}$

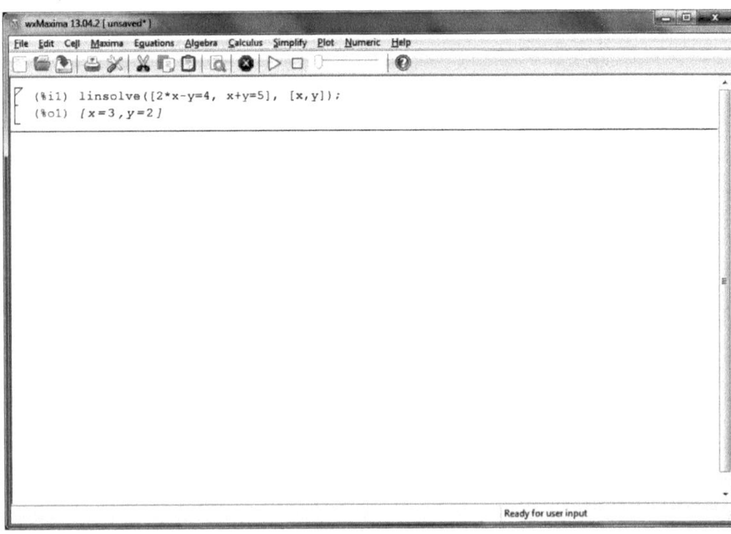

Steps:

1) Go to Maxima
2) Click on Equations → Solve Linear System

3) Choose the number of equations (here 2)

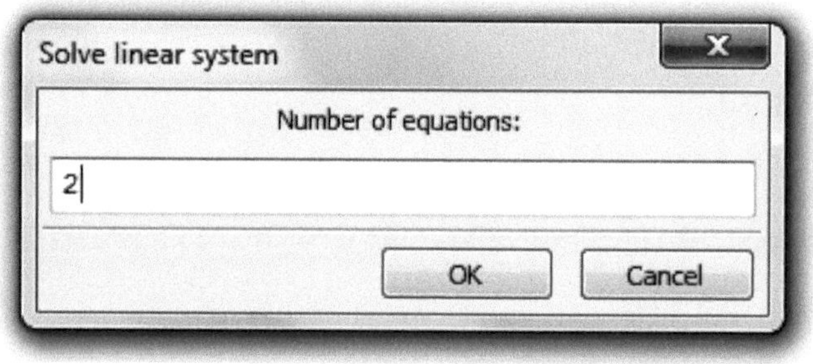

4. Click OK
5. Define your equations and variables

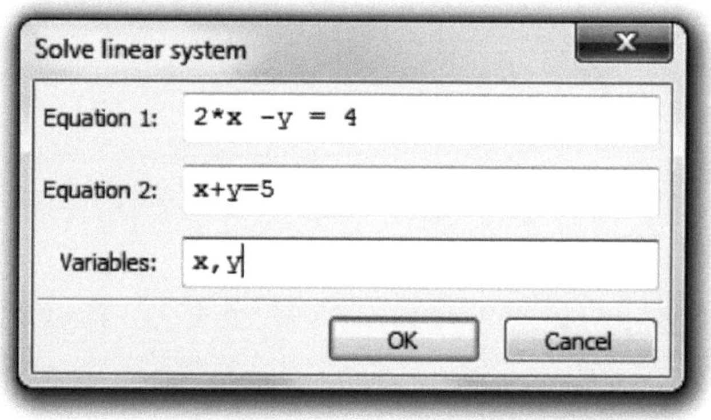

6. Click OK

```
(%i1) linsolve([2*x -y = 4, x+y=5], [x,y]);
(%o1) [x=3,y=2]
```

2. $\begin{cases} 5x+2y=3 \\ 2x+3y=-1 \end{cases}$

```
(%i2) linsolve([5*x+2*y=3, 2*x+3*y =-1], [x,y]);
(%o2) [x=1,y=-1]
```

3. $\begin{cases} 6y-6x=1 \\ 2x+3y=3 \end{cases}$

```
(%i3) linsolve([2*x+3*y=3, 6*y-6*x=1], [x,y]);
```

(%o3) $[x=\frac{1}{2}, y=\frac{2}{3}]$

4. $\begin{cases} 5y=3-2x \\ 3x=2y+1 \end{cases}$

```
(%i4) linsolve([5*y=3-2*x, 3*x=2*y+1], [x,y]);
```

(%o4) $[x=\frac{11}{19}, y=\frac{7}{19}]$

5. $\begin{cases} x=69+6y \\ 3x=4y-45 \end{cases}$

```
(%i5) linsolve([x=69+6*y, 3*x=4*y-45], [x,y]);
(%o5) [x=-39,y=-18]
```

APPLICATION

Solve the following systems using Maxima

1) $\begin{matrix} 4x+2y=9 \\ 3x-2y=10 \end{matrix}$

2) $\begin{cases} 3x-y=7 \\ 2x+3y=1 \end{cases}$

3) $\begin{cases} x+y=10 \\ 2x+y=15 \end{cases}$

4) $\begin{cases} 5x+3y=-19 \\ 8x+3y=-25 \end{cases}$

5) $\begin{cases} 0.45x+0.65y=18.55 \\ x+y=35 \end{cases}$

6) $\begin{cases} 1.8x+1.2y=4 \\ 9x+6y=3 \end{cases}$

7) $\begin{cases} 2x+y=60 \\ x+y=75 \end{cases}$

8) $\begin{cases} 3x+y=-14 \\ 4x+3y=-22 \end{cases}$

9) $\begin{cases} 2x-3y=0 \\ -4x+2y=-8 \end{cases}$

10) $\begin{cases} s+2p=10.25 \\ s+4p=18.75 \end{cases}$

2.10 LEAST COMMON MULTIPLE AND GREAT COMMON DIVISOR

2.10.1 LEAST COMMON MULTIPLE (LCM)

The *lcm* function calculates the least common multiple for two polynomials or integers. This function belongs to the *functs* package, which must be loaded before applying the function. Function *lcm* can be invoked from the *Calculus* menu, however, before using this menu item it is necessary to load the *functs* package. Thus, the first command to enter is:

```
(%i1) load("functs");
(%o1) C:/PROGRA~1/MAXIMA~1.2/share/maxima/5.31.2/share/simplification/functs.mac
```

Find the least common multiple of the following using maxima

a) 2 and 4

```
(%i2) lcm(2,4);
(%o2) 4
```

b) 2 and 3

```
(%i3) lcm(2,3);
(%o3) 6
```

c) 5, 12, 10 and 6

```
(%i4) lcm(5,12,10,6);
(%o4) 60
```

d) $x^2 - 2x$ and $x^3 + 2x^2 - 15x$

We should define first the two polynomials

```
(%i5) p1: x^2 -2*x;
```

(%o5) $x^2 - 2\ x$

```
(%i6) p2: x^3 +2*x^2-15*x;
```

(%o6) $x^3 + 2\ x^2 - 15\ x$

```
(%i7) lcm(p1,p2);
(%o7) (x-3)(x-2)x(x+5)
```

e) $4xy^2$ and $2x^2y^3$

```
(%i8) p1: 4*x*y^2;
(%o8) 4 x y^2

(%i9) p2:2*x^2*y^3;
(%o9) 2 x^2 y^3

(%i10) lcm(p1,p2);
```

f) $16m$, $-12m^2n$ and $8n^2$

```
(%i11) p1: 16*m;
(%o11) 16 m

(%i12) p2:-12*m^2*n;
(%o12) -12 m^2 n

(%i13) p3: 8*n^2;
(%o13) 8 n^2

(%i14) lcm(p1,p2,p3);
(%o14) -48 m^2 n^2
```

g) $x - 1$ and $x + 4$

```
(%i15) p1: x-1;
(%o15) x-1

(%i16) p2:x+4;
(%o16) x+4

(%i17) lcm(p1,p2);
(%o17) (x-1)(x+4)
```

h) y^2 and $y + 3$

```
(%i18) p1:y^2;
(%o18) y^2

(%i19) p2: y+3;
(%o19) y+3

(%i20) lcm(p1,p2);
(%o20) y^2 (y+3)
```

i) $(y - 2)(y + 2)$ and $(y + 2)^2$

```
(%i21) p1:(y-2)*(y+2),
(%o21) (y-2)(y+2)

(%i22) p2:(y+2)^2;
(%o22) (y+2)^2

(%i23) lcm(p1,p2);
(%o23) (y-2)(y+2)^2
```

j) $n^2 - 3n + 2$ and $n^2 - 4$

```
(%i24) p1:n^2-3*n+2;
(%o24) n^2 -3 n+2

(%i25) p2:n^2-4;
(%o25) n^2 -4

(%i26) lcm(p1,p2);
(%o26) (n-2)(n-1)(n+2)
```

k) $x^2 + 9$, $9x^2$ and $x^2 - 6x + 9$

```
(%i27) p1:x^2-9;
(%o27) x^2 -9

(%i29) p2: 9*x^2;
(%o29) 9 x^2

(%i30) p3: x^2-6*x+9;
(%o30) x^2 -6 x+9

(%i31) lcm(p1,p2,p3);
(%o31) 9(x-3)^2 x^2 (x+3)
```

2.10.2 *GREATEST COMMON DIVISOR (GCD)*

The *gcd* function calculates the least common multiple for two polynomials or integers.

Find the greatest common divisor (GCD) of the following using maxima

a) 2 and 4

```
(%i32) gcd(2,4);
(%o32) 2
```

b) 2 and 3

```
(%i33) gcd(2,3);
(%o33) 1
```

c) 6 and 12

```
(%i37) gcd(6,12);
(%o37) 6
```

d) $x^2 - 2x$ and $x^3 + 2x^2 - 15x$

We should define first the two polynomials

```
(%i5) p1: x^2 -2*x;
```

(%o5) $x^2 - 2\,x$

```
(%i6) p2: x^3 +2*x^2-15*x;
```

(%o6) $x^3 + 2\,x^2 - 15\,x$

```
(%i7) lcm(p1,p2);
(%o7) (x-3)(x-2)x(x+5)
```

e) $16m$, $-12m^2n$ and $8n^2$

```
(%i11) p1: 16*m;
(%o11) 16 m

(%i12) p2:-12*m^2*n;
(%o12) -12 m^2 n

(%i13) p3: 8*n^2;
(%o13) 8 n^2

(%i39) gcd(p1,p2,p3);
(%o39) 1
```

f) $x-1$ and $x+4$

```
(%i15) p1: x-1;
(%o15) x-1

(%i16) p2:x+4;
(%o16) x+4

(%i40) gcd(p1,p2);
(%o40) 1
```

g) y^2 and $y^2 + 3$

```
(%i18) p1:y^2;
(%o18) y^2

(%i41) p2: y^2+3;
(%o41) y^2+3

(%i42) gcd(p1,p2);
(%o42) 1
```

h) $(y + 2)$ and $(y + 2)^2$

```
(%i50) p: y+2;
(%o50) y+2

(%i51) p1: (y+2)^2;
(%o51) (y+2)^2

(%i52) gcd(p,p1);
(%o52) y+2
```

i) $n - 2$ and $n^2 - 4$

```
(%i47) p: n-2;
(%o47) n-2

(%i48) p1: n^2-4;
(%o48) n^2-4

(%i49) gcd(p,p1);
(%o49) n-2
```

j) $x^2 - 1$and $x^3 - 1$

```
(%i53) p: x^2-1;
(%o53) x^2 -1

(%i54) p1: x^3-1;
(%o54) x^3 -1

(%i55) gcd(p,p1);
(%o55) x -1
```

APPLICATION

1) Find the LCM of the following
   a) 24 and 42
   b) 36 and 60
   c) 12, 18 and 40
   d) 45, 80 and 120
   e) 36, 153 and 120
   f) 60 and 80
   g) 6, 4 and 10
   h) $-9a^3b$ and $12a^2bc$
   i) $x(x-1$, x^2 and $(x-1)^2$
   j) $5xy$, $15x^2z$ and $10y^2$
   k) $z + 8$ and $z + 2$
   l) x and $x-2$
   m) $x^2 - 1$and $x^2 + 2x + 1$
   n) t, $t^2 - 1$ and $t^2 + 5t - 6$
   o) $8x - 4$ and $6x^2 + x - 2$
   p) $x^3 - y^3$, $x^2 - xy + y^2$ and $x^2 - 2xy + y^2$
2) Find the GCD of the following
   a) 24 and 42
   b) 36 and 60
   c) 12, 18 and 40

d) 45, 80 and 120
e) 36, 153 and 120
f) 60 and 80
g) 6, 4 and 10
h) $-9a^3b$ and $12a^2bc$
i) $x(x-1)$, x^2 and $(x-1)^2$
j) $5xy$, $15x^2z$ and $10y^2$
k) $z + 8$ and $z + 2$
l) x and $x - 2$
m) $x^2 - 1$ and $x^2 + 2x + 1$
n) t, $t^2 - 1$ and $t^2 + 5t - 6$
o) $8x - 4$ and $6x^2 + x - 2$
p) $x^3 - y^3$, $x^2 - xy + y^2$ and $x^2 - 2xy + y^2$

```
(%i6) wxplot3d(x*sin(y)+y*sin(x), [x,-5,5], [y,-5,5])$
```

(%t6)

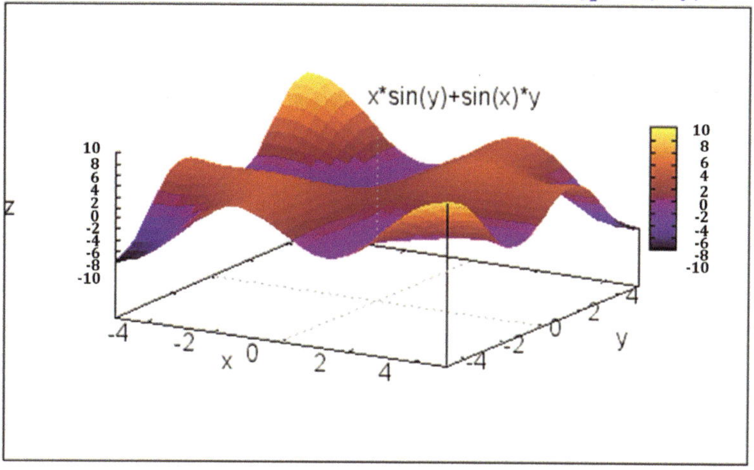

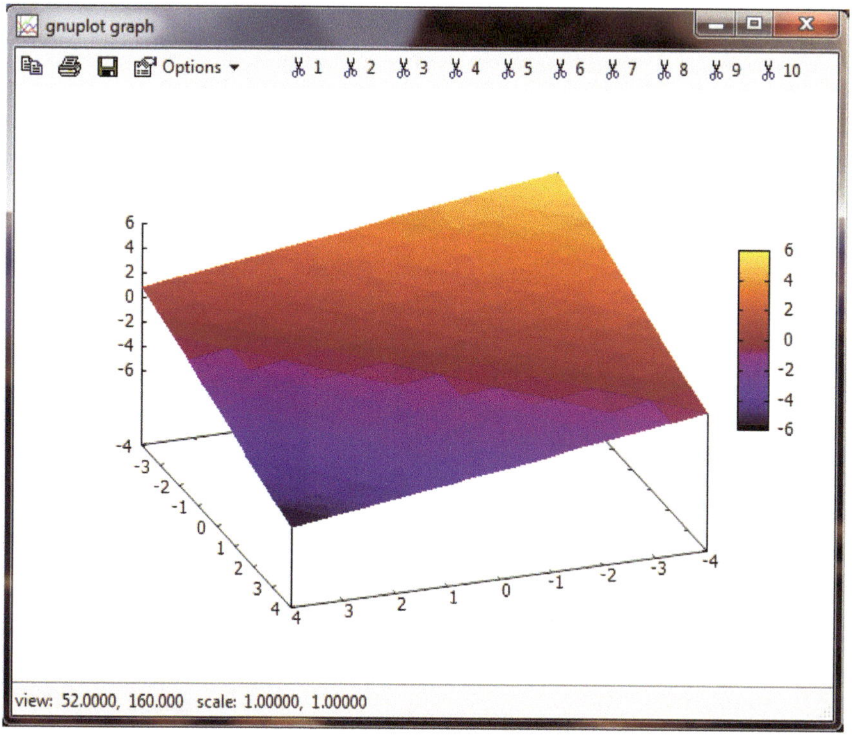

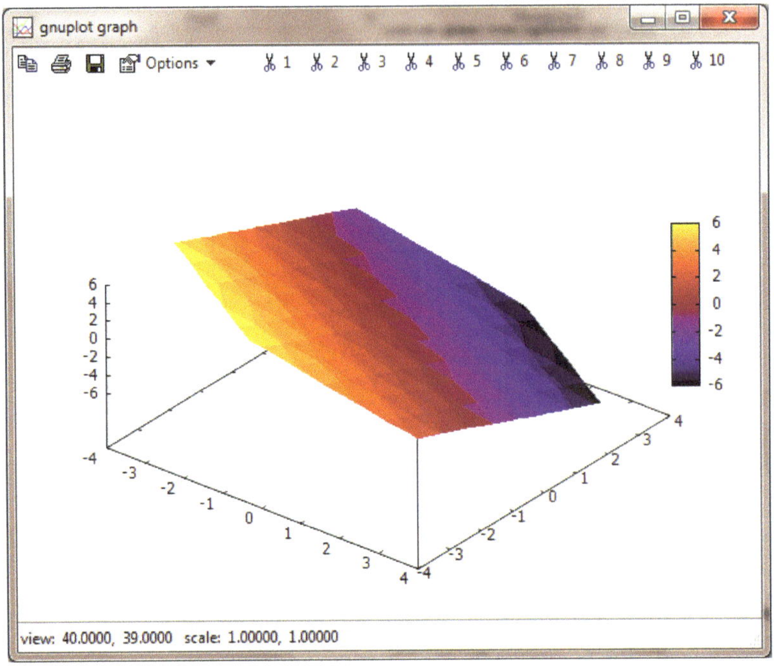
gnuplot graph
Options
1 2 3 4 5 6 7 8 9 10
view: 40.0000, 39.0000 scale: 1.00000, 1.00000

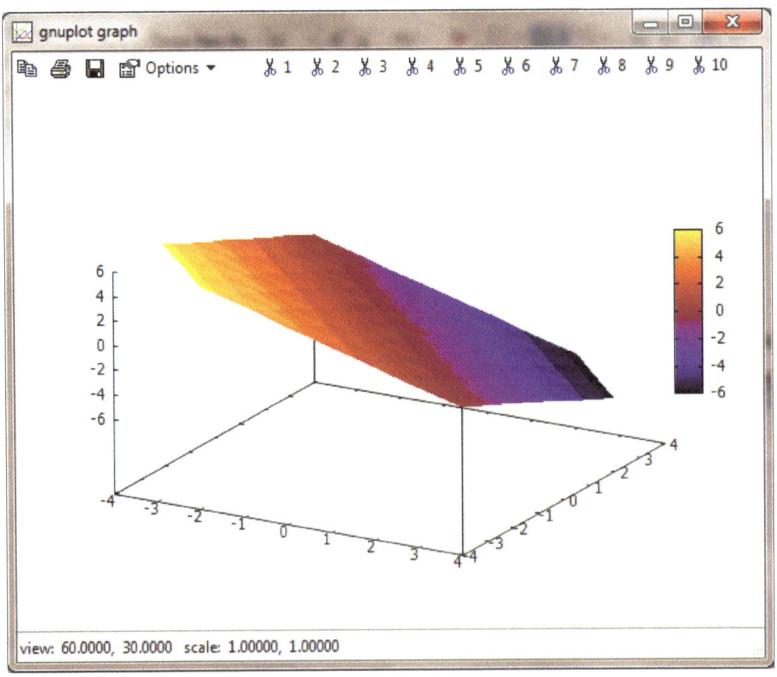
gnuplot graph
Options
1 2 3 4 5 6 7 8 9 10
view: 60.0000, 30.0000 scale: 1.00000, 1.00000

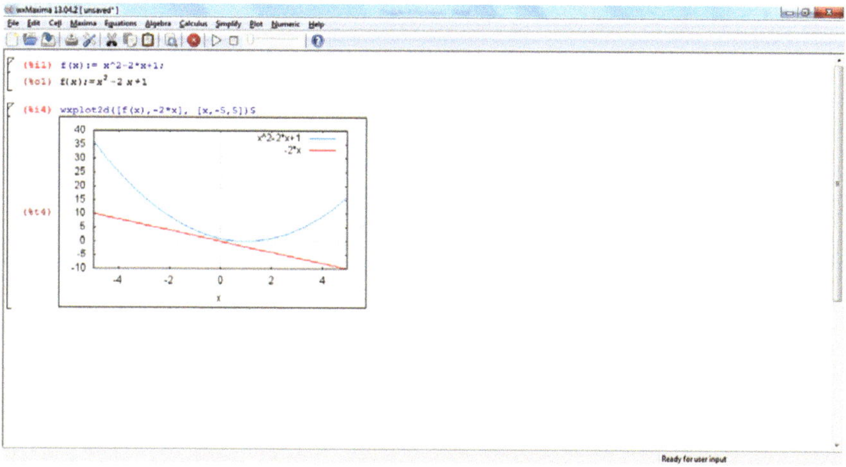


```
(%i5) f(x) := -x^2+5*x;
```

(%o5) $f(x):=-x^2+5\ x$

```
(%i6) wxplot2d([f(x),x], [x,-5,5])$
```

(%t6)

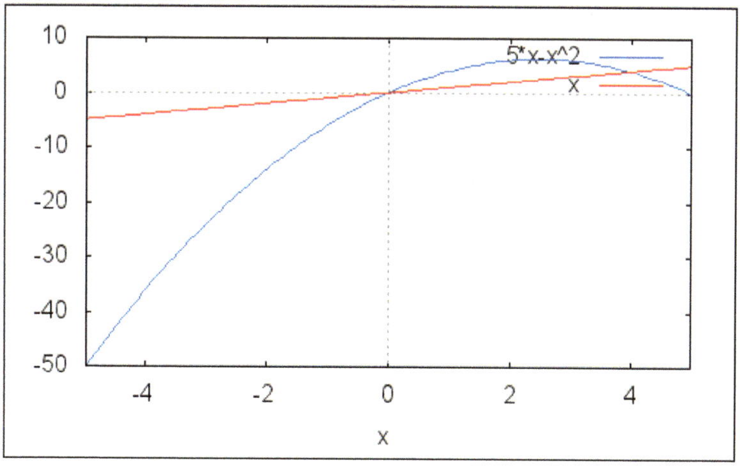

```
(%i7) f(x):=x^2-1;
```

(%o7) $f(x):=x^2-1$

```
(%i8) wxplot2d([f(x),-2*x-1], [x,-5,5])$
```

(%t8)

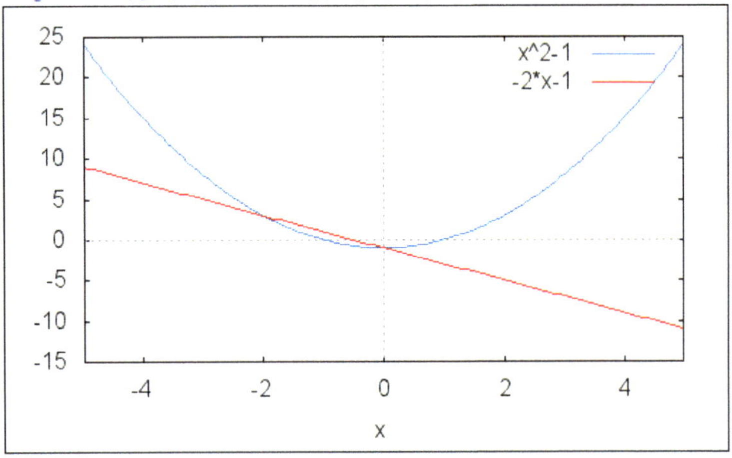


```
(%i11) f(x):= x^2-2;
```

(%o11) $f(x):=x^2-2$

```
(%i12) wxplot2d([f(x),2*x+1], [x,-5,5])$
```

(%t12)

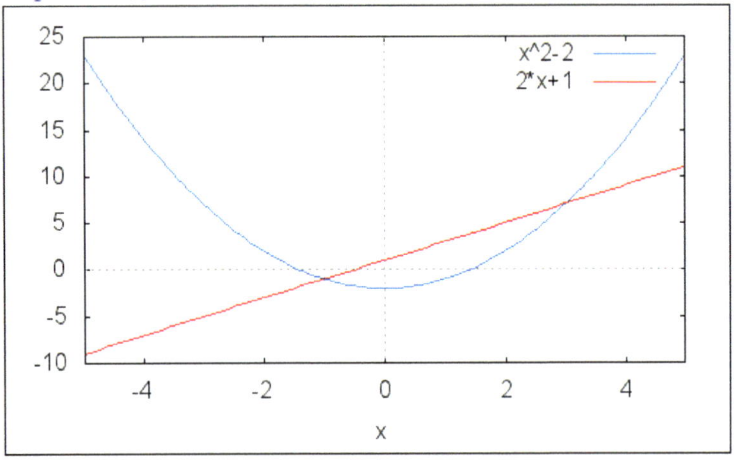

```
(%i13) f(x):= x^2-5*x+7;
```

(%o13) $f(x):=x^2-5\,x+7$

```
(%i14) wxplot2d([f(x),x+5], [x,-5,5])$
```

(%t14)

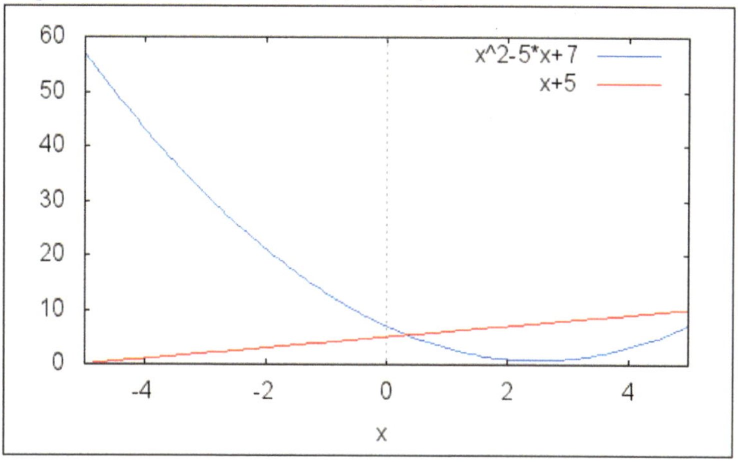

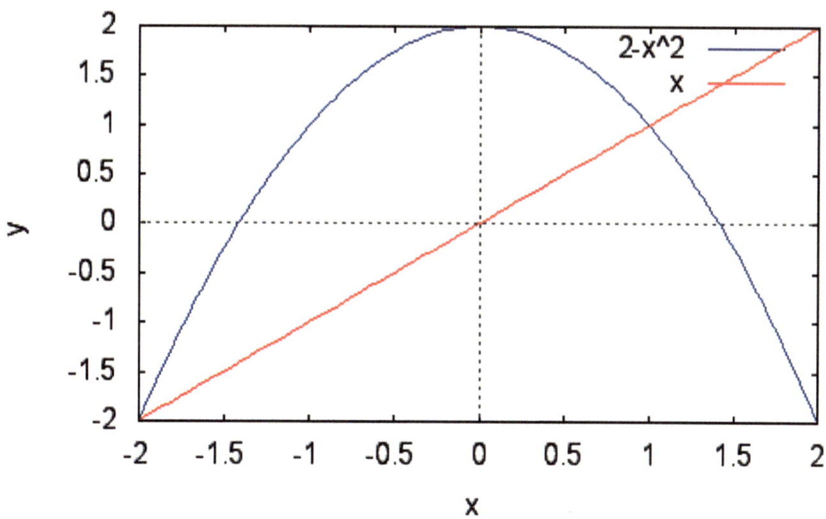

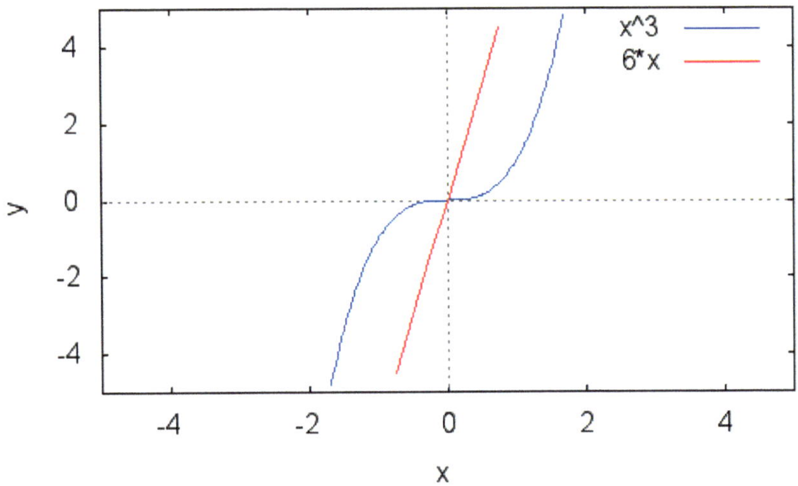
x^3
6*x
y
x
4
2
0
-2
-4

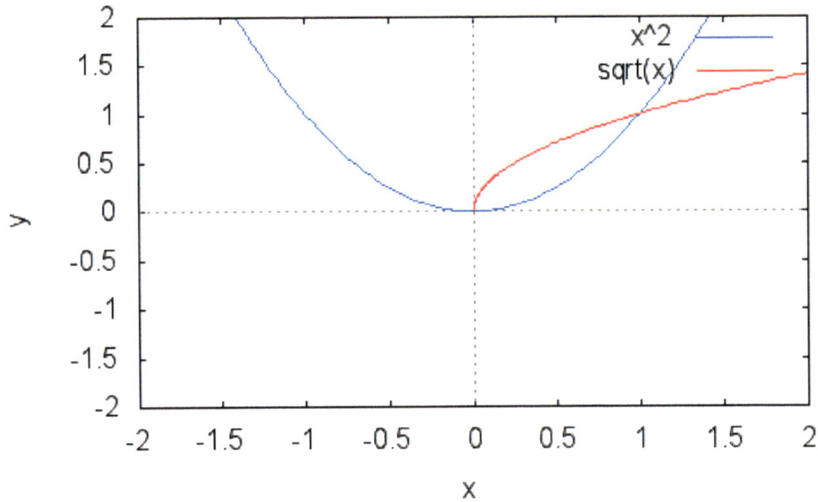
x^2
sqrt(x)
y
x

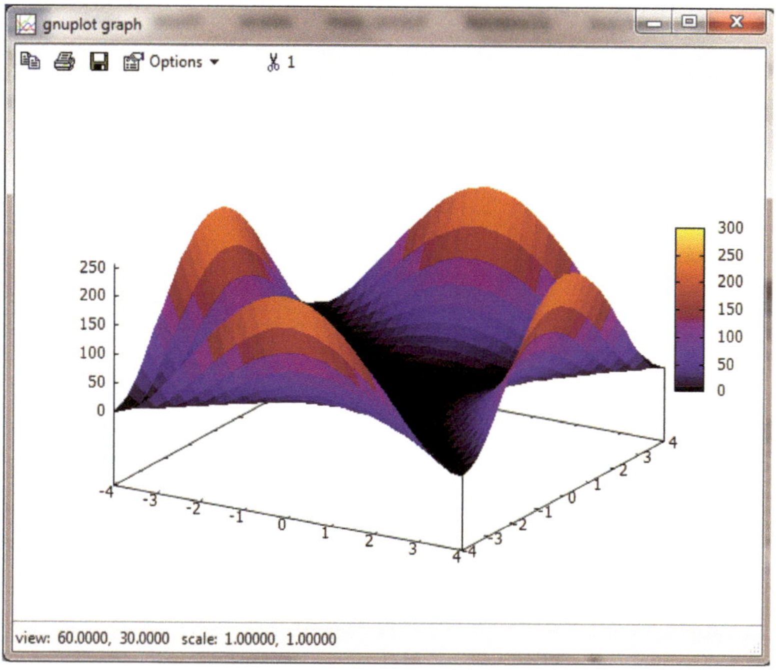
gnuplot graph
Options
1
250
200
150
100
50
0
-4
-3
-2
-1
0
1
2
3
4
4
3
2
1
0
-1
-2
-3
-4
300
250
200
150
100
50
0
view: 60.0000, 30.0000 scale: 1.00000, 1.00000

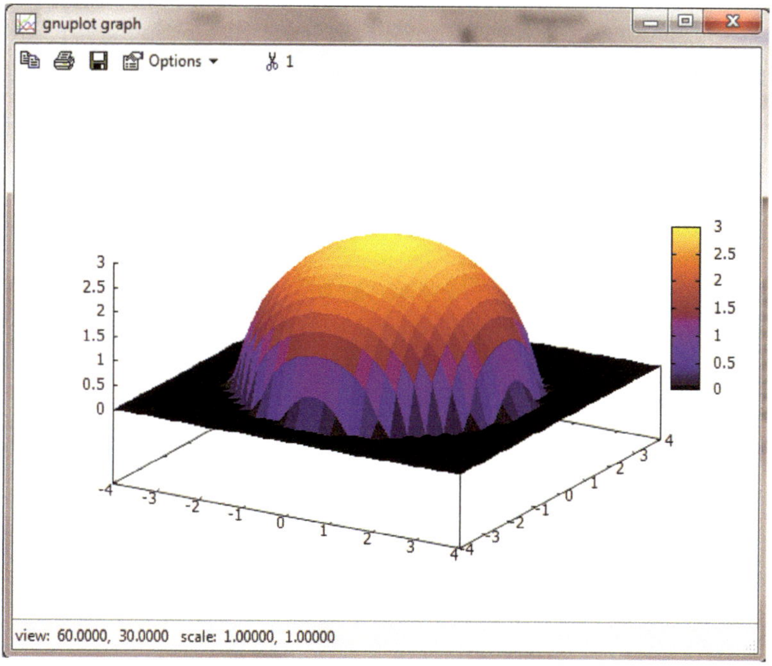
gnuplot graph
Options
1
3
2.5
2
1.5
1
0.5
0
-4
-3
-2
-1
0
1
2
3
4
4
3
2
1
0
-1
-2
-3
-4
3
2.5
2
1.5
1
0.5
0
view: 60.0000, 30.0000 scale: 1.00000, 1.00000

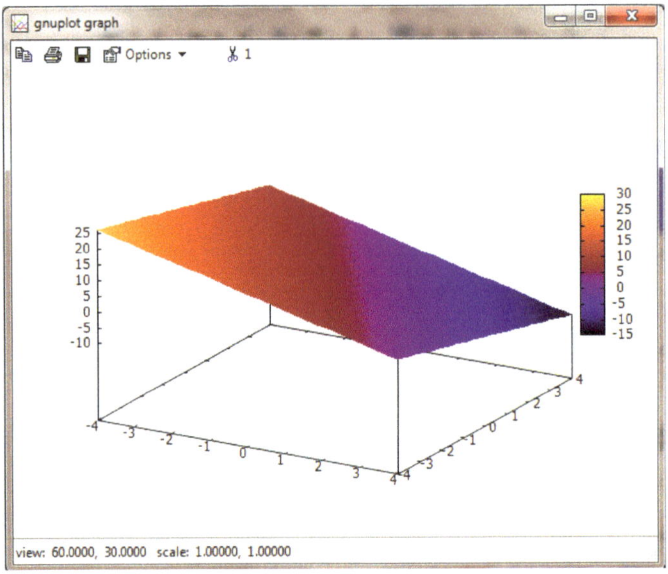
gnuplot graph
Options
1
view: 60.0000, 30.0000 scale: 1.00000, 1.00000

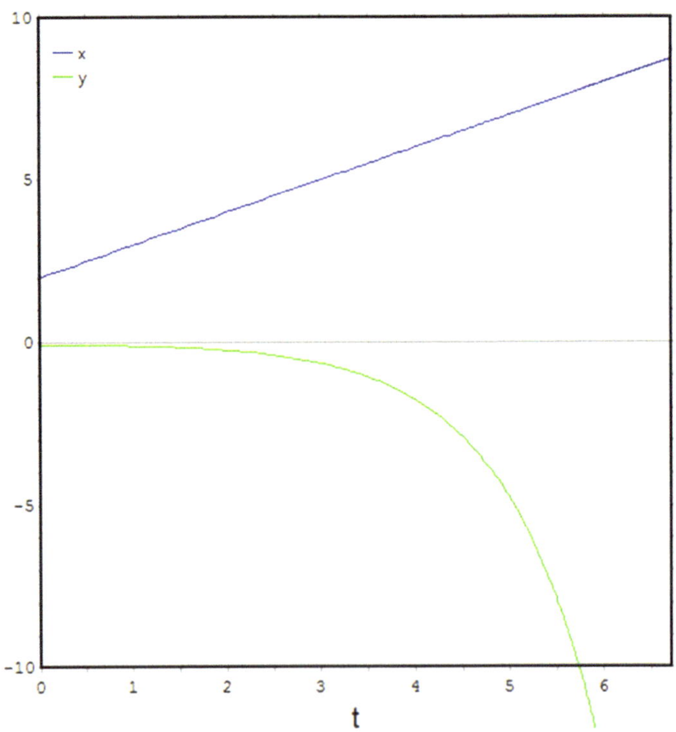
x
y
t

CHAPTER 3

CALCULUS

3.1 COMPLEX NUMBERS

1. Define the following complex numbers using Maxima

   (1) $z = 4 - \frac{1}{2}i$

NB: The equal sign "=" should be replaced by two dots ":"

"i" is defined as constant but "i" here is a complex constant function since

$i^2 = 1$, so we have to write "%i"

Steps:

1) Go to Maxima
2) Type in the command z: 4 – ½*%i
3) Press on Shift Enter

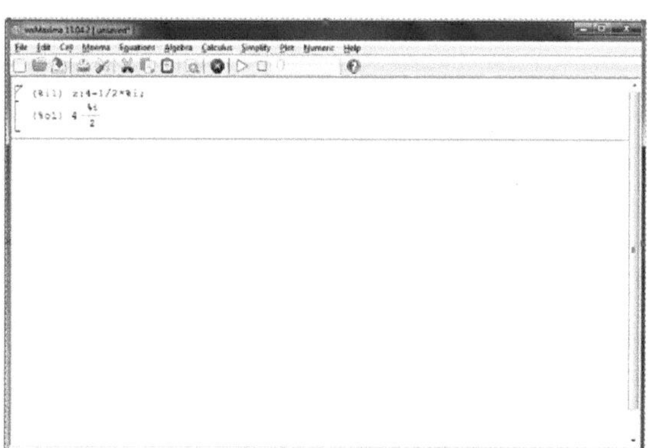


```
(%i1) z:4-1/2*%i;
```

$$(\%o1)\quad 4 - \frac{\%i}{2}$$

(2) $w = 9 + \frac{5}{2}i$

```
(%i2) w:9+5/2*%i;
       5 %i
(%o2)  ---- +9
        2
```

(3) $r = 1 - 2i$

```
(%i3) r:1-2*%i;
(%o3) 1 -2 %i
```

(4) $t = 8 - 3i$

```
(%i4) t:8-3*%i;
(%o4) 8 -3 %i
```

(5) $b = 5 + 2i$

```
(%i5) b:5+2*%i;
(%o5) 2 %i +5
```

(6) $j = i$

```
(%i6) j:%i;
(%o6) %i
```

(7) $q = 3 + 6i$

```
(%i7) q:3+6*%i;
(%o7) 6 %i +3
```

(8) $u = -2 + i$

```
(%i8) u:-2+%i;
(%o8) %i-2
```

2. Find the conjugate of the following complex numbers using Maxima

1) $z = 4 - \frac{1}{2}i$

Steps:

1) Go to Maxima
2) Define the complex number by typing: $\left(z:4-\frac{1}{2}*\%i\right)$
3) Press Shift Enter
4) To find the conjugate we just need to write in the command "**conjugate (z)**"

conjugate (z)

5) Press Shift Enter to execute

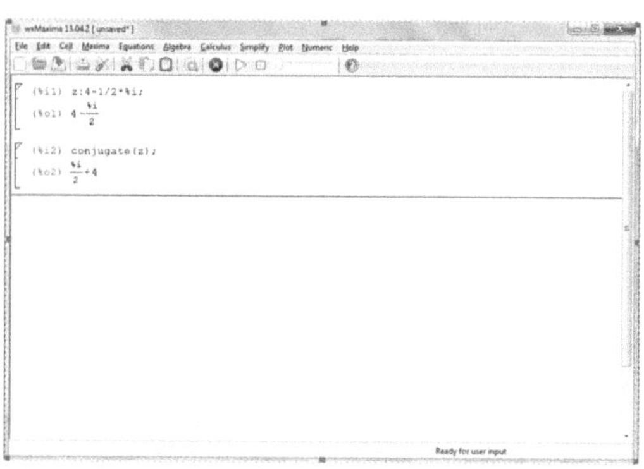


```
(%i9) conjugate(z);
       %i
(%o9)  -- + 4
       2
```

2) $w = 9 + \frac{5}{2}i$

```
(%i11) conjugate(w);
(%o11) 9 - 5 %i / 2
```

3) $r = 1 - 2i$

```
(%i12) conjugate(r);
(%o12) 2 %i + 1
```

4) $t = 8 - 3i$

```
(%i13) conjugate(t);
(%o13) 3 %i + 8
```

5) $b = 5 + 2i$

```
(%i14) conjugate(b);
(%o14) 5 - 2 %i
```

6) $j = i$

```
(%i15) conjugate(j);
(%o15) - %i
```

7) $q = 3 + 6i$

```
(%i16) conjugate(q);
(%o16) 3 - 6 %i
```

8) $u = -2 + i$

```
(%i17) conjugate(u);
(%o17) -%i -2
```

3. Find the modulus of the following complex numbers using Maxima

1) $z = 4 - \frac{1}{2}i$

Steps:

1) Go to Maxima
2) Define the complex number by typing: $\left(z:4 - \frac{1}{2} * \%i\right)$
3) Press Shift Enter
4) To find the modulus we just have to write in the command the function "**cabs (z)**"
5) Press Shift Enter to execute

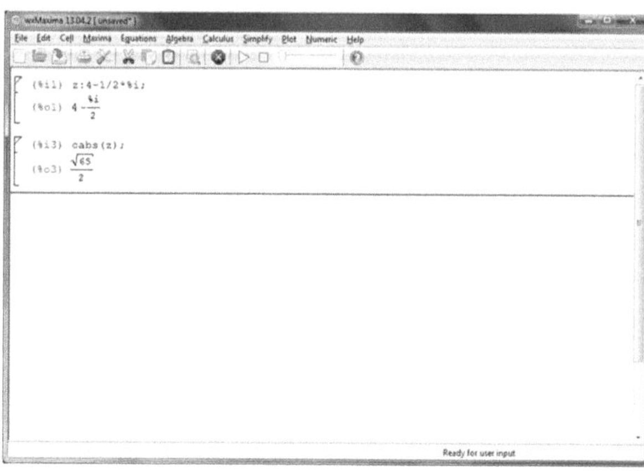


```
(%i10) cabs(z);
```

(%o10) $\frac{\sqrt{65}}{2}$

2) $w = 9 + \frac{5}{2}i$

```
(%i18) cabs(w);
```

(%o18) $\frac{\sqrt{349}}{2}$

3) $r = 1 - 2i$

```
(%i19) cabs(r);
```

(%o19) $\sqrt{5}$

4) $t = 8 - 3i$

```
(%i20) cabs(t);
```

(%o20) $\sqrt{73}$

5) $b = 5 + 2i$

```
(%i21) cabs(b);
```

(%o21) $\sqrt{29}$

6) $j = i$

```
(%i22) cabs(j);
(%o22) 1
```

7) $q = 3 + 6i$

```
(%i23) cabs(q);
```

(%o23) $3\sqrt{5}$

8) $u = -2 + i$

```
(%i24) cabs(u);
(%o24) √5
```

4. Evaluate the expression and write your answer in the form $a + bi$ using Maxima

1) $\left(4-\frac{1}{2}i\right)+\left(9+\frac{5}{2}i\right)$

Steps:

1) Go to Maxima
2) Define the complex number by typing: $\left(z:4-\frac{1}{2}*\%i\right)$
3) Press Shift Enter
4) Define the second complex number by typing: $\left(w:9+\frac{5}{2}*\%i\right)$
5) Press Shift Enter
6) Press on “z+w”
7) Press Shift Enter

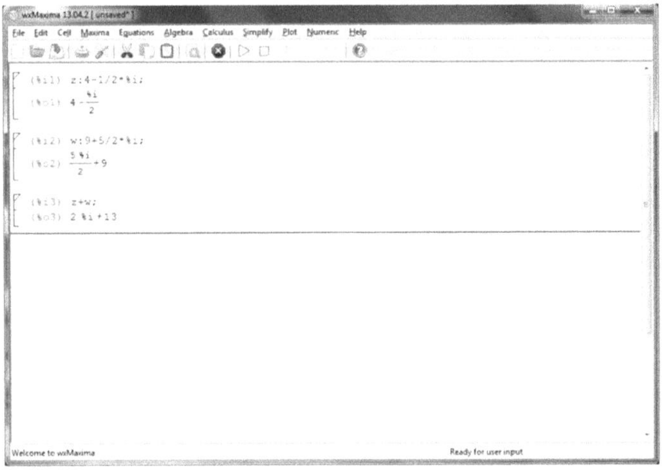


```
(%i25) z+w;
(%o25) 2 %i +13
```

2) $\left(4-\frac{1}{2}i\right)+\left(9+\frac{5}{2}i\right)$

Steps:

1) Go to Maxima
2) Define the complex number by typing: $\left(z:4-\frac{1}{2}*\%i\right)$
3) Press Shift Enter
4) Define the second complex number by typing: $\left(w:9+\frac{5}{2}*\%i\right)$
5) Press Shift Enter
6) Press on “z*w”
7) Press Shift Enter
8) Go to Simplify → Simplify Expression

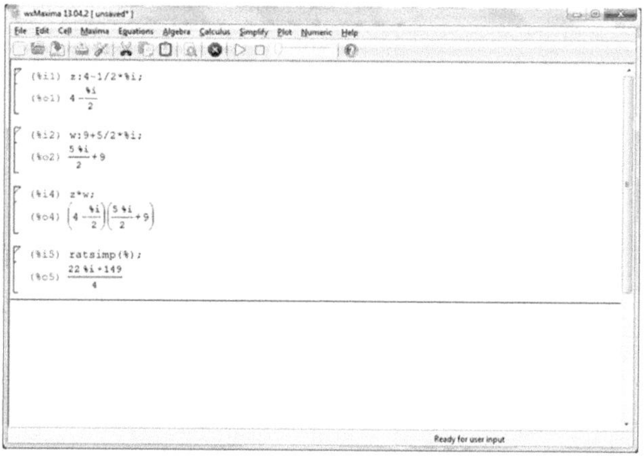


```
(%i33) z*w;
(%o33) (4 - %i/2)(5 %i/2 + 9)

(%i34) ratsimp(%);
(%o34) (22 %i + 149)/4
```

3) i^3

```
(%i26) %i^3;
(%o26) -%i
```

4) i^{100}

```
(%i28) %i^100;
(%o28) 1
```

5) (3 + 6i)(–2 + *i*)

```
(%i29) q+u;
(%o29) 7 %i + 1
```

6) (3 + 6*i*)(–2 + *i*)

```
(%i30) q-u;
(%o30) 5 %i +5
```

7) (3 + 6*i*)(–2 + *i*)

```
(%i35) q*u;
(%o35) (%i -2)(6 %i +3)
```

```
(%i36) ratsimp(%);
(%o36) -9 %i -12
```

8) (3 + 6*i*)/(–2 + *i*)

Steps:

1) Go to Maxima
2) Define the complex number by typing: (*q*: 3–6 *% *i*)
3) Press Shift Enter
4) Define the second complex number by typing: (*u*: –2 + %*i*)
5) Press Shift Enter
6) Press on "q/u"
7) Press Shift Enter
8) Go to Simplify → Complex Simplification → Convert to Rectform

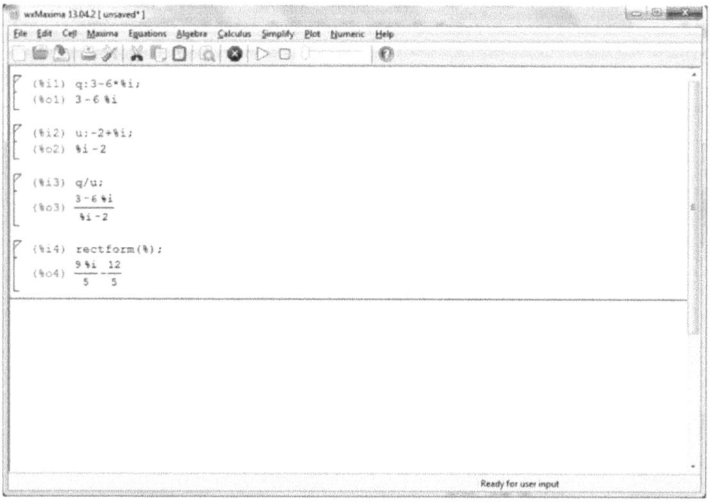


```
(%i41) q/u;
        6 %i + 3
(%o41)  --------
         %i - 2

(%i42) rectform(%);
(%o42) -3 %i
```

9) $(1-2i)+(8-3i)$

```
(%i31) r+t;
(%o31) 9 - 5 %i
```

10) $\left(4-\dfrac{1}{2}i\right)+\left(9+\dfrac{5}{2}i\right)$

```
(%i32) z-w;
(%o32) -3 %i - 5
```

11) $\left(4-\frac{1}{2}i\right)/\left(9+\frac{5}{2}i\right)$

```
(%i43) z/w;
```

$$(\%o43)\quad \frac{4-\frac{\%i}{2}}{\frac{5\,\%i}{2}+9}$$

```
(%i44) rectform(%);
```

$$(\%o44)\quad \frac{139}{349}-\frac{58\,\%i}{349}$$

5. Find all solutions of the equation using Maxima

1) $4x^2 + 9 = 0$

Steps:

1) Go to Maxima
2) Click on Equations → Solve
3) Define your equation

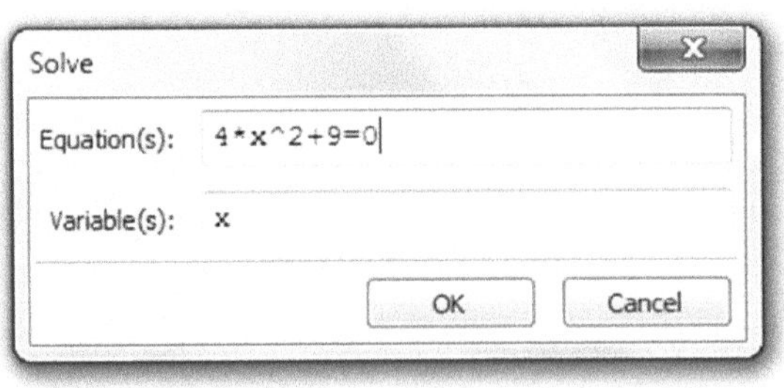

4) Click on OK

```
(%i45) solve([4*x^2+9=0], [x]);
```

(%o45) $[x=-\frac{3\,\%i}{2}, x=\frac{3\,\%i}{2}]$

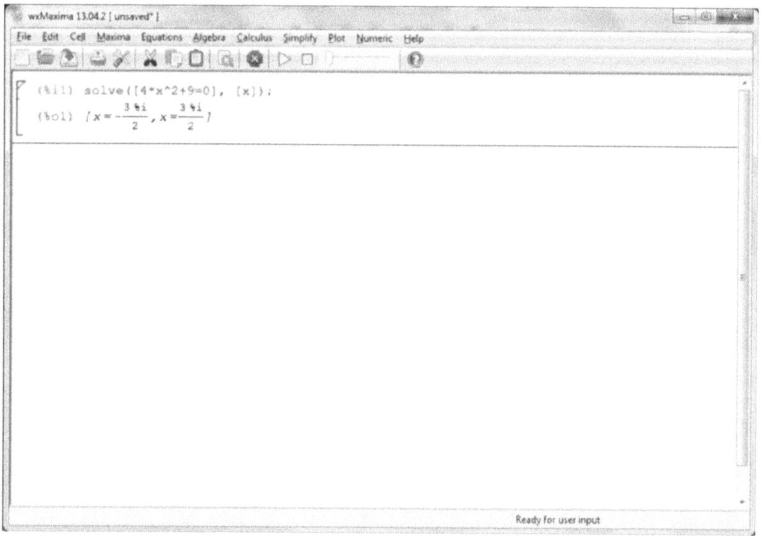

2) $2x^2 - 2x + 1 = 0$

```
(%i46) solve([2*x^2-2*x+1 = 0], [x]);
```

(%o46) $[x=-\frac{\%i-1}{2}, x=\frac{\%i+1}{2}]$

3) $z^2 + z + 2 = 0$

```
(%i47) solve([z^2+z+2=0], [x]);
(%o47) []
```

4) $3x^2 - 7x + 5 = 0$

```
(%i49) solve([3*x^2-7*x+5=0], [x]);
```

$$(\%o49)\quad [x=-\frac{\sqrt{11}\,\%i-7}{6},x=\frac{\sqrt{11}\,\%i+7}{6}]$$

5) $-4x^2 + 5x - 3 = 0$

```
(%i50) solve([-4*x^2+5*x-3=0], [x]);
```

$$(\%o50)\quad [x=-\frac{\sqrt{23}\,\%i-5}{8},x=\frac{\sqrt{23}\,\%i+5}{8}]$$

6) $2x^2 - 8x + 26 = 0$

```
(%i51) solve([2*x^2-8*x+26=0], [x]);
(%o51) [x=2-3 %i,x=3 %i+2]
```

7) $3x^2 - 12x + 12 = 0$

```
(%i52) solve([3*x^2-12*x+12=0], [x]);
(%o52) [x=2]
```

8) $2x^2 + 4 = 0$

```
(%i53) solve([2*x^2+4=0], [x]);
```

$$(\%o53)\quad [x=-\sqrt{2}\,\%i,x=\sqrt{2}\,\%i]$$

6. Find the Real and the Imaginary part of the following complex numbers using Maxima

1) $z = 4 - \frac{1}{2}i$

Real part

Steps:

1) Go to Maxima
2) Define your complex number
3) Go to Simplify → Complex Simplification → Get Real Part

```
(%i54) realpart(z:4-1/2*%i);
(%o54) 4
```

Imaginary part

Steps:

1) Go to Maxima
2) Define your complex number
3) Go to Simplify → Complex Simplification → Get Imaginary Part

```
(%i55) imagpart(z:4-1/2*%i);
```

(%o55) $-\frac{1}{2}$

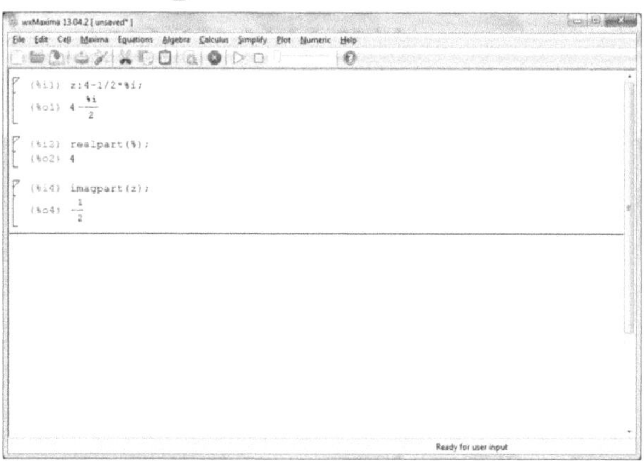

2) $w = 9 + \frac{5}{2}i$

```
(%i56) realpart(w:9+5/2*%i);
(%o56) 9

(%i57) imagpart(w:9+5/2*%i);
        5
(%o57)  -
        2
```

3) $r = 1 - 2i$

```
(%i58) realpart(r:1-2*%i);
(%o58) 1

(%i59) imagpart(r:1-2*%i);
(%o59) -2
```

4) $t = 8 - 3i$

```
(%i60) realpart(t:8-3*%i);
(%o60) 8

(%i61) imagpart(t:8-3*%i);
(%o61) -3
```

5) $q = 3 + 6i$

```
(%i62) realpart(q:3+6*%i);
(%o62) 3

(%i63) imagpart(q:3+6*%i);
(%o63) 6
```

6) $u = -2 + i$

```
(%i64) realpart(u:-2+%i);
(%o64) -2

(%i65) imagpart(u:-2+%i);
(%o65) 1
```

7) $j = i$

```
(%i72) realpart(%);
(%o72) 0

(%i70) imagpart(j:%i);
(%o70) 1
```

APPLICATION

1. Define the following complex numbers using Maxima
   1) $z = 8 + 10i$
   2) $w = -3 + 5i$
   3) $t = 7 - 6i$
   4) $q = -3 + 8i$
   5) $r = -2 - 9i$
   6) $u = 4 + 3i$
   7) $v = -4 - 3i$
   8) $k = 2 - 12i$
   9) $s = 2 + i$
   10) $f = -5 - 3i$

2. Find the conjugate of the following complex numbers using Maxima
   1) $z = 8 + 10i$
   2) $w = -3 + 5i$

3) $t = 7–6i$
4) $q = –3 + 8i$
5) $r = –2 – 9i$
6) $u = 4 + 3i$
7) $v = –4 – 3i$
8) $k = 2 – 12i$
9) $s = 2 + i$
10) $f = –5 – 3i$

3. Find the modulus of the following complex numbers using Maxima
   1) $z = 8 + 10i$
   2) $w = –3 + 5i$
   3) $t = 7 – 6i$
   4) $q = –3 + 8i$
   5) $r = –2 – 9i$
   6) $u = 4 + 3i$
   7) $v = –4 – 3i$
   8) $k = 2 – 12i$
   9) $s = 2 + i$
   10) $f = –5 – 3i$

4. Evaluate the expression and write your answer in the form $a + bi$ using Maxima
   1) $(8 + 10i) + (–3 + 5i)$
   2) $(8 + 10i) – (–3 + 5i)$
   3) $(8 + 10i)(–3 + 5i)$
   4) $(8 + 10i)/(–3 + 5i)$
   5) $(7 – 6i)(4 + 3i)$
   6) $(7 – 6i) + (4 + 3i)$
   7) $(7 – 6i) – (4 + 3i)$
   8) $(7 – 6i)/(4 + 3i)$

5. Find all solutions of the equation using Maxima
   1) $x^2 – 4x + 13 = 0$
   2) $2x^2 + 16x + 33 = 0$
   3) $x^2 + 1 = 0$
   4) $x^2 – 8x + 17 = 0$
   5) $x^2 – 2x + 10 = 0$

6) $-16t^2 + 34t - 20 = 0$
7) $3x^2 - 5x + 3 = 0$
8) $6x + 2x^2 + 9 = 0$

6. Find the Real and the Imaginary part of the following complex numbers using Maxima
   1) $z = 8 + 10i$
   2) $w = -3 + 5i$
   3) $t = 7 - 6i$
   4) $q = -3 + 8i$
   5) $r = -2 - 9i$
   6) $u = 4 + 3i$
   7) $v = -4 - 3i$
   8) $k = 2 - 12i$
   9) $s = 2 + i$
   10) $f = -5 - 3$

3.2 VECTORS

1. Define the following vectors using Maxima

   1) $u = [-1, 2]$

Steps:
1) Go to Maxima
2) Write the vector name "u:[–1,2]"
3) Press Shift Enter

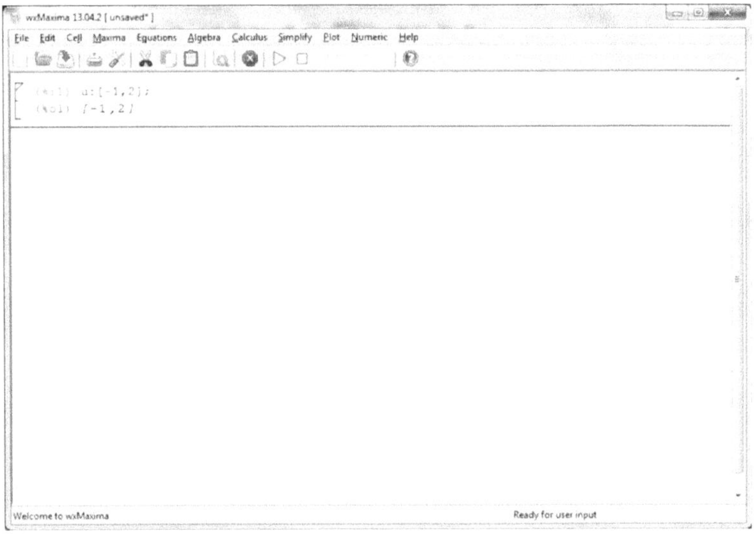

2) $A = [-4, -1]$

```
(%i2) A:[-4,-1];
(%o2) [-4,-1]
```

3) $B = [2,2]$

```
(%i3) B:[2,2];
(%o3) [2,2]
```

4) $C = [-1,4]$

```
(%i4) C:[-1,4];
(%o4) [-1,4]
```

5) $m = [-3,7]$

```
(%i5) m:[-3,7];
(%o5) [-3,7]
```

2. Find the addition and subtraction of the following vectors using Maxima

1) $u = [-1,2]$ $\quad$ $A = [-4, -1]$

Addition:

Steps:

1) Go to Maxima
2) Define the vectors
3) Write this command "u+A"

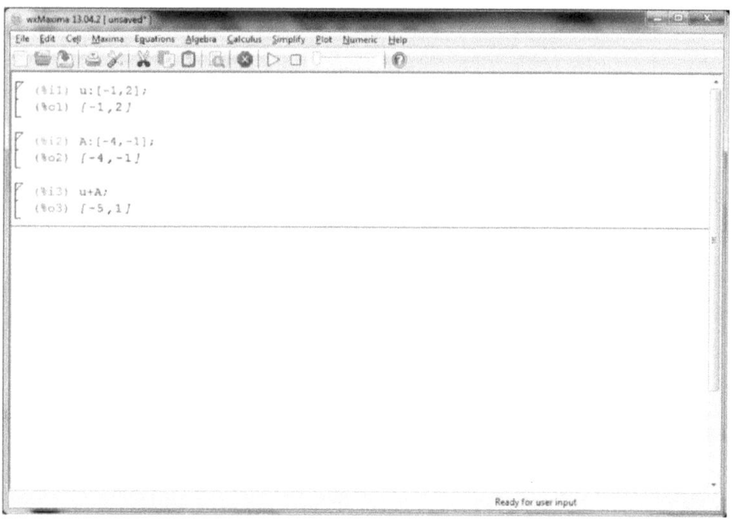

Subtraction:

Steps:

1) Go to Maxima
2) Define the vectors
3) Write this command "u-A"

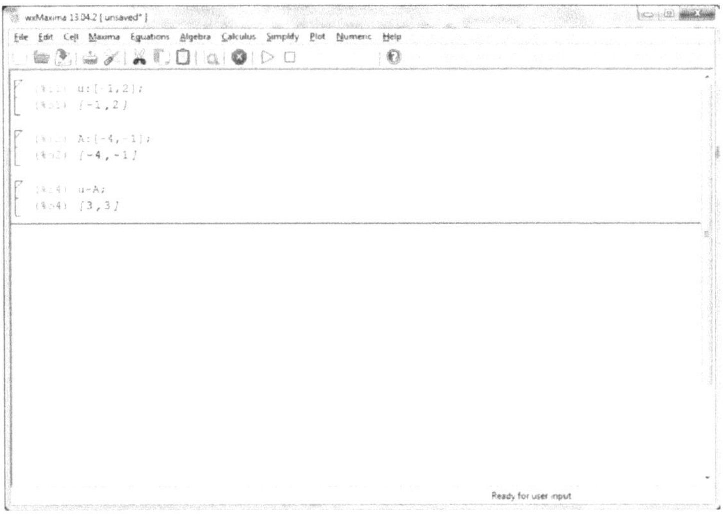

2) $B = [2,2]$ $C = [-1,4]$

```
(%i5) B:[2,2];
(%o5) [2,2]

(%i6) C:[-1,4];
(%o6) [-1,4]

(%i7) B+C;
(%o7) [1,6]

(%i8) B-C;
(%o8) [3,-2]
```

3) $m = [-3,7]$ $n = [4,5]$

```
(%i9) m:[-3,7];
(%o9) [-3,7]

(%i10) n:[4,5];
(%o10) [4,5]

(%i11) m+n;
(%o11) [1,12]

(%i12) m-n;
(%o12) [-7,2]
```

4) $r = [0,1]$ $t = [-3,2]$

```
(%i13) r:[0,1];
(%o13) [0,1]

(%i14) t:[-3,2];
(%o14) [-3,2]

(%i15) r+t;
(%o15) [-3,3]

(%i16) r-t;
(%o16) [3,-1]
```

5) $q = [5,7]$ $s = [-1; -4]$

```
(%i17) q:[5,7];
(%o17) [5,7]

(%i18) s:[-1,-4];
(%o18) [-1,-4]

(%i19) q+s;
(%o19) [4,3]

(%i20) q-s;
(%o20) [6,11]
```

3. Given the two vectors $a = [-2,8]$ and $b = 3i + 5j$. Find the following:

    1) $a \div 2$

Steps:

1) Go to Maxima
2) Define the two vectors
3) Type the command "a/2"
4) Press Shift Enter

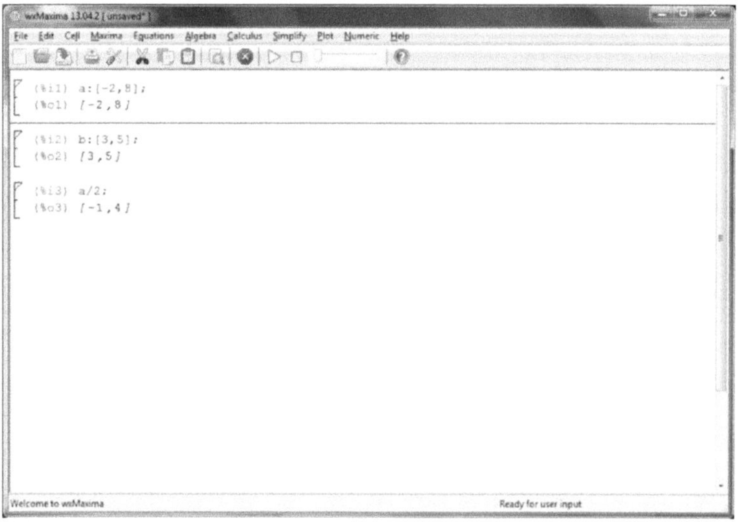

2) $3b$

```
(%i4) 3*b;
(%o4) [9,15]
```

3) $5a$

```
(%i5) 5*a;
(%o5) [-10,40]
```

4) $b \div 2$

```
(%i6) b/2;
```

(%o6) $[\frac{3}{2},\frac{5}{2}]$

5) $10a$

```
(%i7) 10*a;
(%o7) [-20,80]
```

4. Find the length of the following vectors using Maxima

1) u = [–8, 2]

Steps:

1) Go to Maxima
2) Define the vector
3) Write the following function "**sqrt(u.u)**"
4) Press Shift Enter

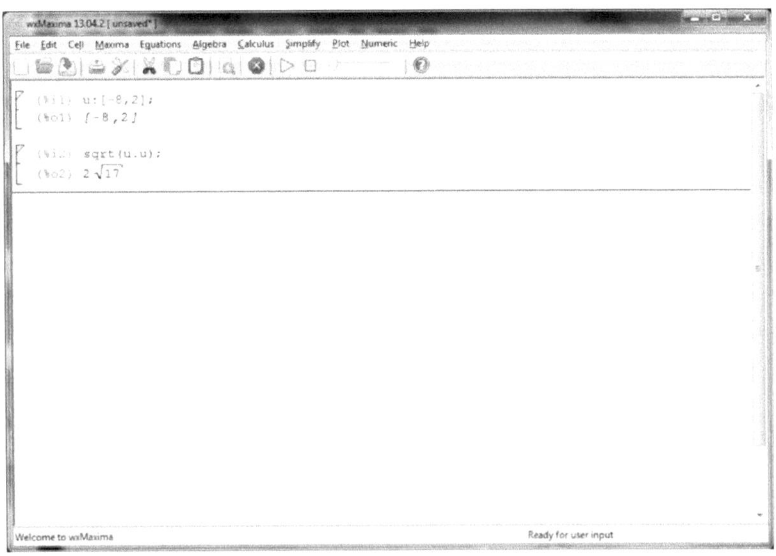

2) v = [3,4]

```
(%i5) v:[2,3];
(%o5) [2,3]

(%i6) sqrt(v.v);
(%o6) √13
```

3) $t = [-3,2]$

```
(%i10) t:[-3,2];
(%o10) [-3,2]

(%i11) sqrt(t.t);
(%o11) √13
```

4) $a = [-4, -6]$

```
(%i12) a:[-4,-6];
(%o12) [-4,-6]

(%i13) sqrt(a.a);
(%o13) 2 √13
```

5) $b = [9,3]$

```
(%i14) b:[9,3];
(%o14) [9,3]

(%i15) sqrt(b.b);
(%o15) 3 √10
```

5. Find the unit vector of the following vectors using Maxima

1) $u = [-8,2]$

Steps:

1) Go to Maxima
2) Define the vector
3) Write the following "**load("eigen")**" (you just need to write it once)
4) Press Shift Enter
5) Write "uvect(u)"
6) Press Shift Enter

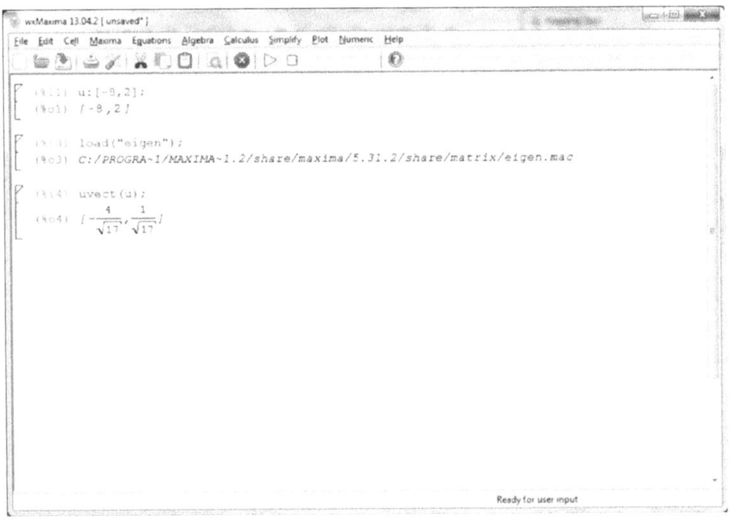

2) $v = [3,4]$

```
(%i5) v:[2,3];
(%o5) [2,3]

(%i7) load("eigen");
(%o7) C:/PROGRA~1/MAXIMA~1.2/share/maxima/5.31.2/share/matrix/eigen.mac

(%i8) uvect(v);
```

$$(\%o8)\ [\frac{2}{\sqrt{13}}, \frac{3}{\sqrt{13}}]$$

3) $t = [-3,2]$

```
(%i10) t:[-3,2];
(%o10) [-3,2]
```

```
(%i16) uvect(t);
```

(%o16) $[-\frac{3}{\sqrt{13}}, \frac{2}{\sqrt{13}}]$

4) $a = [-4, -6]$

```
(%i12) a:[-4,-6];
(%o12) [-4,-6]
```

```
(%i17) uvect(a);
```

(%o17) $[-\frac{2}{\sqrt{13}}, -\frac{3}{\sqrt{13}}]$

5) $b = [9,3]$

```
(%i14) b:[9,3];
(%o14) [9,3]
```

```
(%i18) uvect(b);
```

(%o18) $[\frac{3}{\sqrt{10}}, \frac{1}{\sqrt{10}}]$

6. Find the dot product of the two vectors

1) $u = [-1,2]$ $\quad A = [-4, -1]$

Steps:

1) Go to Maxima
2) Define the two vectors
3) Write "a.b"
4) Press Shift Enter

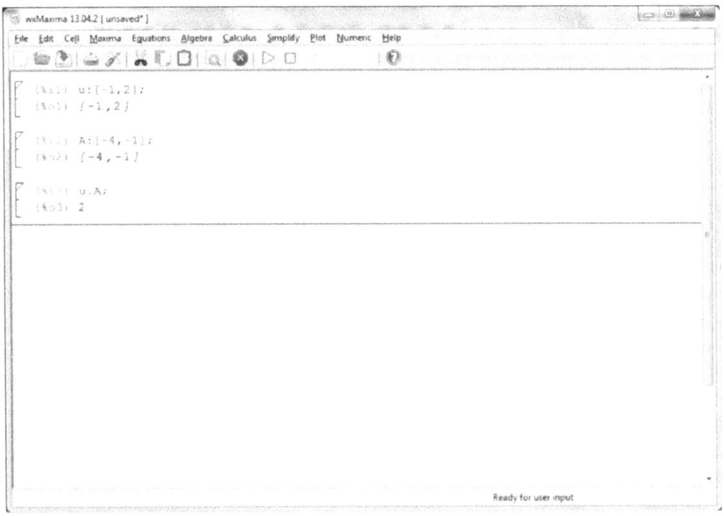

2) $B = [2,2]$ $C = [-1,4]$

```
(%i4) B:[2,2];
(%o4) [2,2]

(%i5) C:[-1,4];
(%o5) [-1,4]

(%i6) B.C;
(%o6) 6
```

3) $m = [-3,7]$ $n = [4,5]$

```
(%i7) m:[-3,7];
(%o7) [-3,7]

(%i8) n:[4,5];
(%o8) [4,5]

(%i9) m.n;
(%o9) 23
```

4) $r = [0,1]$ $t = [-3,2]$

```
(%i10) r:[0,1];
(%o10) [0,1]

(%i11) t:[-3,2];
(%o11) [-3,2]

(%i12) r.t;
(%o12) 2
```

5) $q = [5,7]$ $s = [-1; -4]$

```
(%i13) q:[5,7];
(%o13) [5,7]

(%i14) s:[-1,-4];
(%o14) [-1,-4]

(%i15) q.s;
(%o15) -33
```

7. Find the cross product using Maxima

   1) $a = < 1,2,3 >$ $\qquad b = < 2, 3, 4 >$

Steps:

1) Go to Maxima
2) **Load("vect")$**
3) [1, 2, 3] ~ [2, 3, 4]
4) Press Shift Enter
5) Type **express(%)**
6) Press Shit Enter

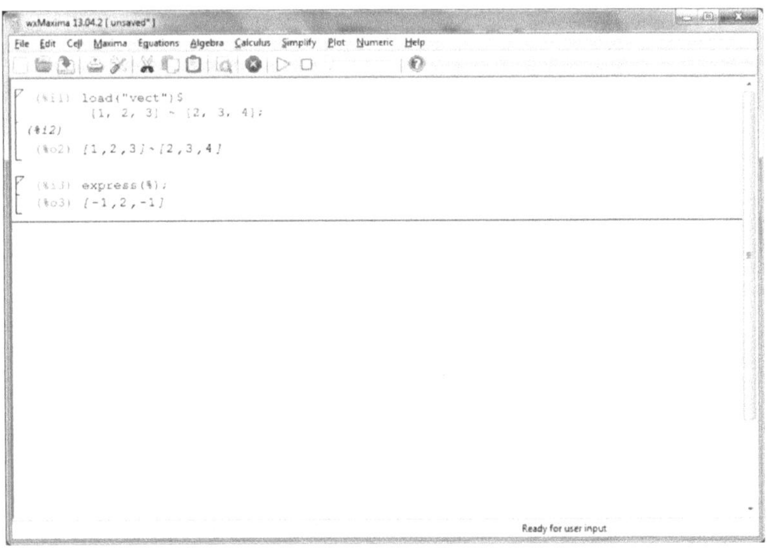


```
(%i1) load("vect")$
        [1, 2, 3] ~ [2, 3, 4];
(%i2)
(%o2) [1,2,3]~[2,3,4]

(%i3) express(%);
(%o3) [-1,2,-1]
```

2) $a = <1,3,4>$ $b = <2,7,-5>$

```
(%i1) load("vect")$
      [1,3,4]~[2,7,-5];
(%o2) [1,3,4]~[2,7,-5]

(%i3) express(%);
(%o3) [-43,13,1]
```

3) $a = <-3,1,-7$? $b = <0,-5,-5>$

```
(%i1) load("vect")$
      [-3,1,-7]~[0,-5,-5];
(%o2) [-3,1,-7] ~ [0,-5,-5]

(%i3) express(%);
(%o3) [-40,-15,15]
```

4) $a = <6,0,-2>$

```
(%i4) kill(all);
(%o0) done

(%i1) load("vect")$
      [6,0,-2]~[0,8,0];
(%o2) -[0,8,0] ~ [6,0,-2]

(%i3) express(%);
(%o3) [16,0,48]
```

5) $a = < 1,1,-1 >$

```
(%i4) kill(all);
(%o0) done

(%i1) load("vect")$
      [1,1,-1]~[2,4,6];
(%o2) [1,1,-1] ~ [2,4,6]

(%i3) express(%);
(%o3) [10,-8,2]
```

8. Find the vector projection of the following vectors

a) $a = < 1,2,3 >$

Steps:

1) Go to Maxima
2) Define the two vectors
3) Type: "$\left(\dfrac{a \cdot b}{a \cdot a} \times a\right)$"
4) Press Shift Enter

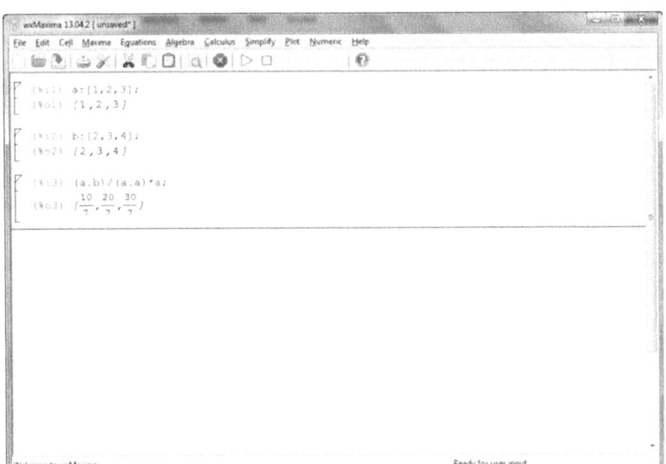

```
(%i1) a:[1,2,3];
(%o1) [1,2,3]

(%i2) b:[2,3,4];
(%o2) [2,3,4]

(%i3) (a.b)/(a.a)*a;
```

(%o3) $[\frac{10}{7},\frac{20}{7},\frac{30}{7}]$

b) $a = < 1,3,4 >$ $b = < 2,7,-5 >$

```
(%i4) a:[1,3,4];
(%o4) [1,3,4]

(%i5) b:[2,7,-5];
(%o5) [2,7,-5]

(%i6) (a.b)/(a.a)*a;
```

(%o6) $[\frac{3}{26},\frac{9}{26},\frac{6}{13}]$

c) $a = < -3,1,-7 >$ $b = < 0,-5,-5,-5 >$

```
(%i7) a:[-3,1,-7];
(%o7) [-3,1,-7]

(%i8) b:[0,-5,-5];
(%o8) [0,-5,-5]

(%i9) (a.b)/(a.a)*a;
```

(%o9) $[-\frac{90}{59},\frac{30}{59},-\frac{210}{59}]$

d) $a = <6,0,-2>$ $b = <0,8,0>$

```
(%i10) a:[6,0,-2];
(%o10) [6,0,-2]

(%i11) b:[0,8,0];
(%o11) [0,8,0]

(%i12) (a.b)/(a.a)*a;
(%o12) [0,0,0]
```

e) $a = <1,1,-1>$ $b = <2,4,6>$

```
(%i13) a:[1,1,-1];
(%o13) [1,1,-1]

(%i14) b:[2,4,6];
(%o14) [2,4,6]

(%i16) (a.b)/(a.a)*a;
(%o16) [0,0,0]
```

Equations of lines and planes

1) Plot the plane with equation $3x + 4y + 5z = 0$

Steps:

1) Go to Maxima
2) Define the plane: "P:3*x+4*y+5*z=0"
3) Load(draw)
4) Press Shift Enter
5) Draw3d(enhanced3d = true, implicit(P, x,–4,4, y,–4,4, z,–6,6))
6) Press Shift Enter

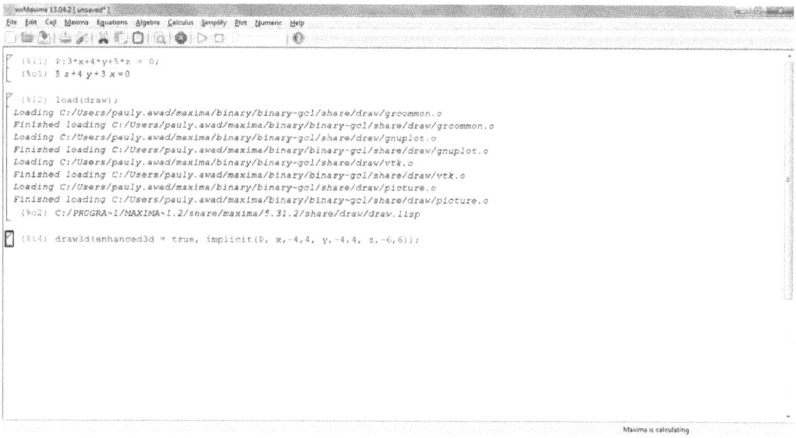

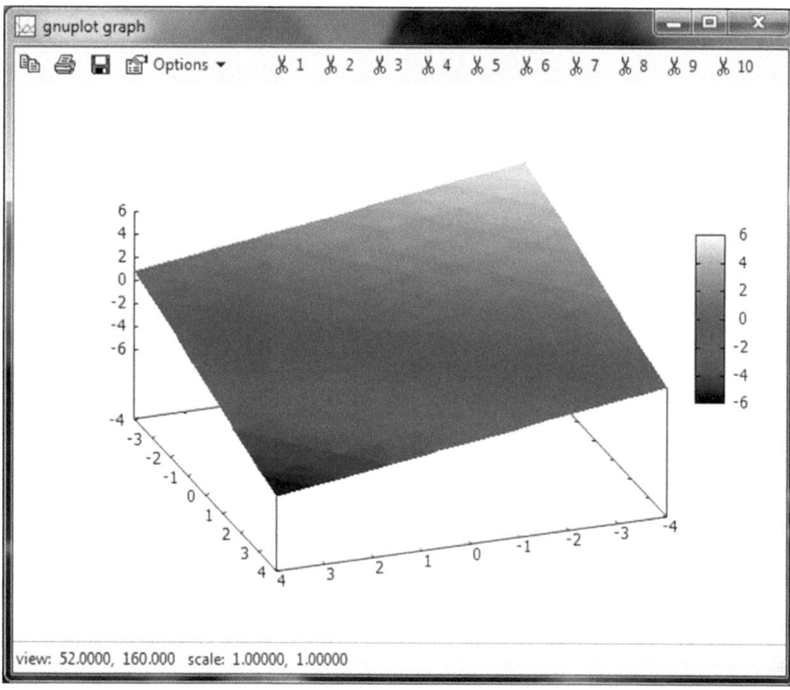

(See color insert.)

NB: We can rotate the graph using Maxima

2) Plot the plane with equation $6x + 10y + 7z = 0$

```
(%i5) P:6*x+10*y+7*z = 0;
(%o5) 7 z+10 y+6 x=0

(%i6) load(draw);
(%o6) C:/PROGRA~1/MAXIMA~1.2/share/maxima/5.31.2/share/draw/draw.lisp

(%i7) draw3d(enhanced3d = true, implicit(P, x,-4,4, y,-4,4, z,-6,6));
```

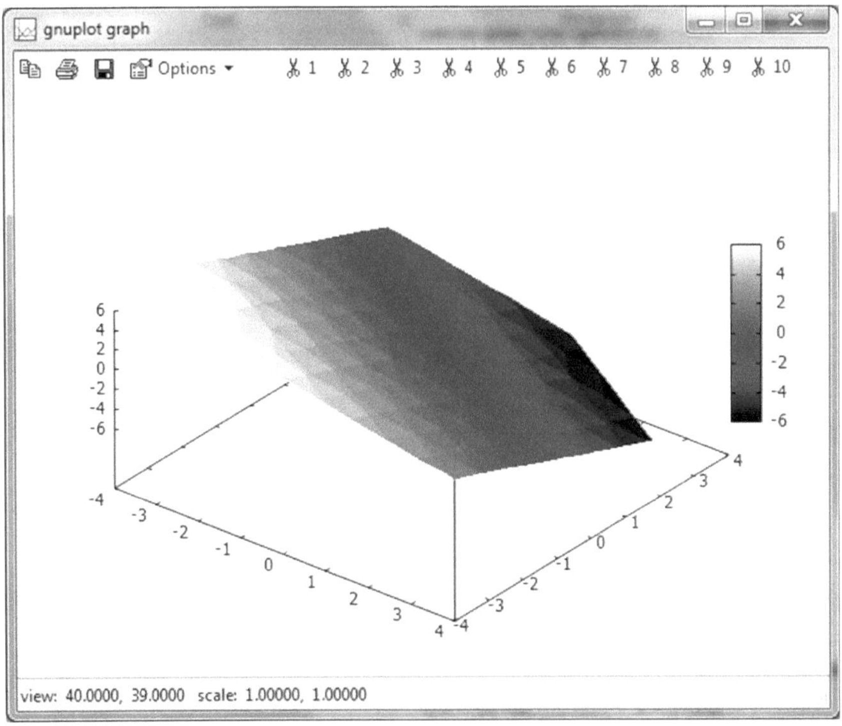

(See color insert.)

3) Plot the plane with equation $x + y + z = 0$

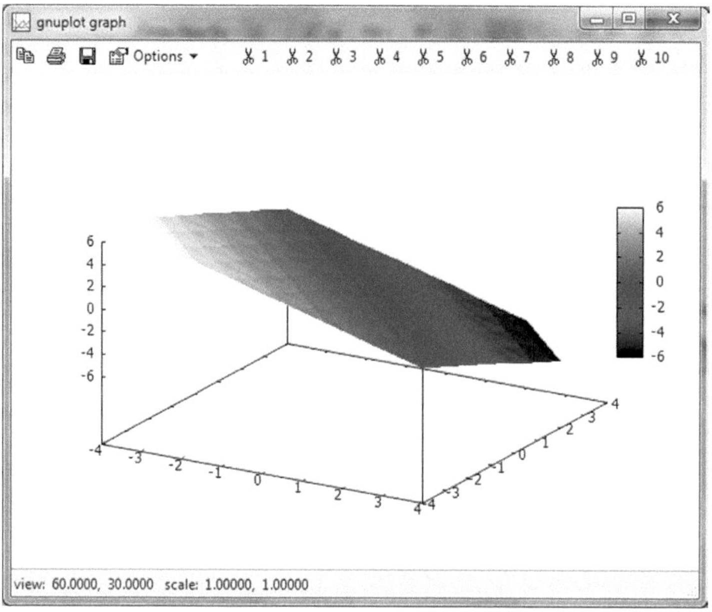

(See color insert.)

APPLICATION

1. Define the following vectors using Maxima
   1) $u = [3, -2]$
   2) $v = [-2,5]$
   3) $K = [0, -6]$
   4) $l = [4, -7]$
   5) $a = [-4,8]$
2. Find the addition and subtraction of the following vectors using Maxima
   1) $u = [3, -2]$ $v = [-2, 5]$
   2) $K = [0, -6]$ $l = [4, -7]$
   3) $a = [-4,8]$ $b = [3, -8]$
   4) $c = [3,3]$ $d = [0, -4]$
   5) $e = [6,3]$ $f = [-5,7]$
3. Given the two vectors $a = [-6,4]$ and $\mathrm{b} = 2i - 3j$. Find the following:
   1) $3a$
   2) $3b$
   3) $a \div 6$

   4) $b \div 3$
   5) $7a$
4. Find the length of the following vectors using Maxima
   1) $u = [3, -2]$
   2) $v = [-2,5]$
   3) $K = [0, -6]$
   4) $l = [4, -7]$
   5) $a = [-4,8]$
5. Find the unit vector of the following vectors using Maxima
   1) $u = [3, -2]$
   2) $v = [-2,5]$
   3) $K = [0, -6]$
   4) $l = [4, -7]$
   5) $a = [-4,8]$
6. Find the dot product of the two vectors
   1) $u = [3, -2]$ $v = [-2, 5]$
   2) $K = [0, -6]$ $l = [4, -7]$
   3) $a = [-4,8]$ $b = [3, -8]$
   4) $c = [3,3]$ $d = [0, -4]$
   5) $e = [6,3]$ $f = [-5,7]$
7. Find the cross product of the two vectors
   1) $a = < 1,3,-2 >$ $b = < -1,0,5 >$
   2) $a = < -1,4,-2 >$ $b = < -1,7,-2 >$
   3) $a = < 1,3,-1 >$ $b = < -1/5,0,1/2 >$
   4) $a = < 2,-1,3 >$ $b = < 4,2,1 >$
   5) $\mathrm{a} = < 1,0,1 >$ $b = < 2,1,-1 >$
8. Find the vector projection of the following vectors
   1) $a = < 1,3,-2 >$ $b = < -1,0,5 >$
   2) $\mathrm{a} = < -1,4,-2 >$ $\mathrm{b} = < -1,7,-2 >$
   3) $a = < 1, 3, -1 >$ $\mathrm{b} = < -1/5,0,1 >$
   4) $a = < 2,-1,3 >$ $b = < 4,2,1 >$
   5) $a = < 1,0,1 >$ $b = < 2,1,-1 >$
9. Plot the plane with equation $2x + 5y - z = 0$
10. Plot the plane with equation $x - 5y + 3z = 0$

3.3 FUNCTIONS

1. Define and plot the following functions using Maxima
   1) $f(x) = 3x^2 + 1$

Steps:

1) Go to Maxima
2) To define the function we have to write "f(x):=3*x^2+1"
3) Press Shift Enter
4) To plot the function go to Plot → Plot 2d
5) In the expression(s) box write "f(x)" or "f(x):=3*x^2+1"
6) Click on ok

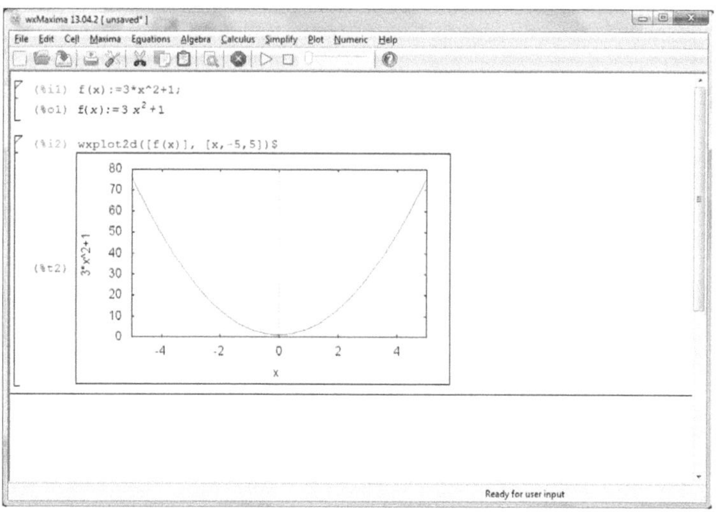

2) $g(x) = -x^2 - 1$

```
(%i3) g(x):= -x^2-1;
(%o3) g(x):=-x^2-1
```

```
(%i4) wxplot2d([g(x)], [x,-5,5])$
```

(%t4)

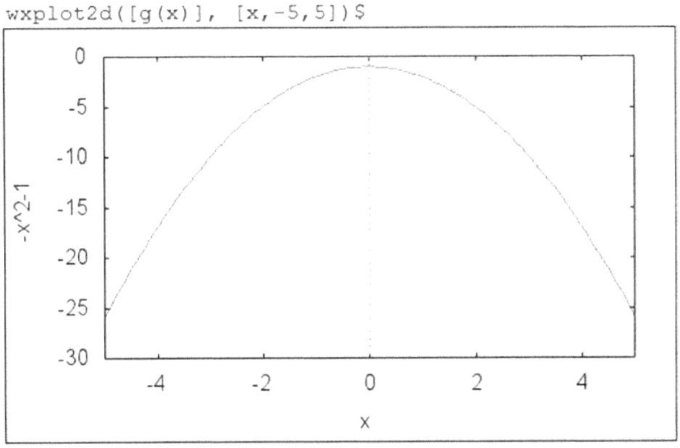

3) $h(x) = x^2 + 2x + 2$

```
(%i5) h(x):= x^2+2*x+2;
(%o5) h(x):=x^2+2 x+2
```

```
(%i7) wxplot2d([h(x)], [x,-5,5])$
```

(%t7)

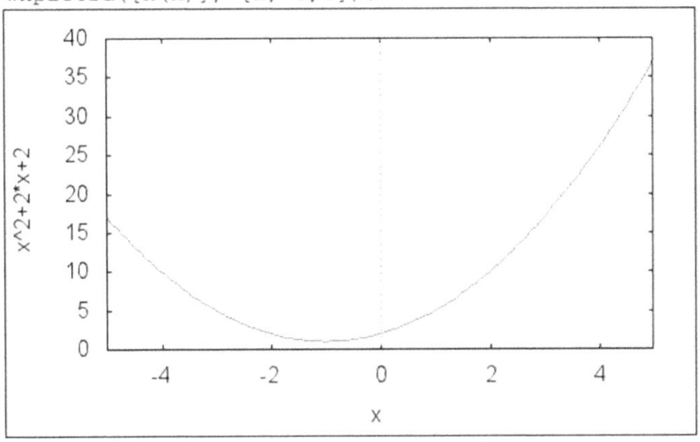

4) $f(x) = x^2 + 1$

```
(%i8) f(x):= x^2+1;
(%o8) f(x):=x^2+1

(%i9) wxplot2d([f(x)], [x,-5,5])$
```

(%t9)

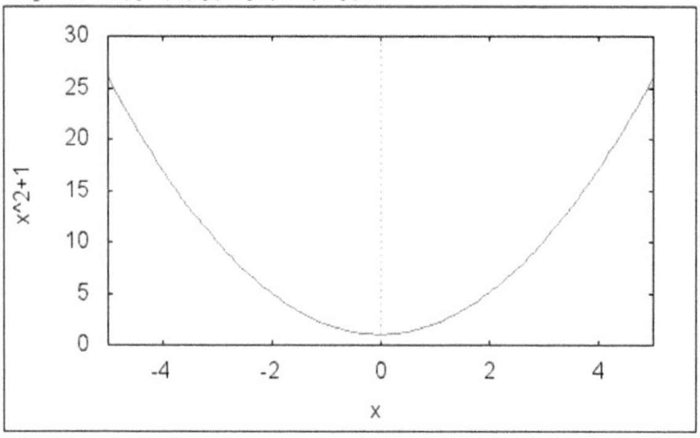

5) $r(x) = x^3 + 1$

```
(%i10) r(x):=x^3+1;
(%o10) r(x):=x^3+1

(%i11) wxplot2d([r(x)], [x,-5,5])$
```

(%t11)

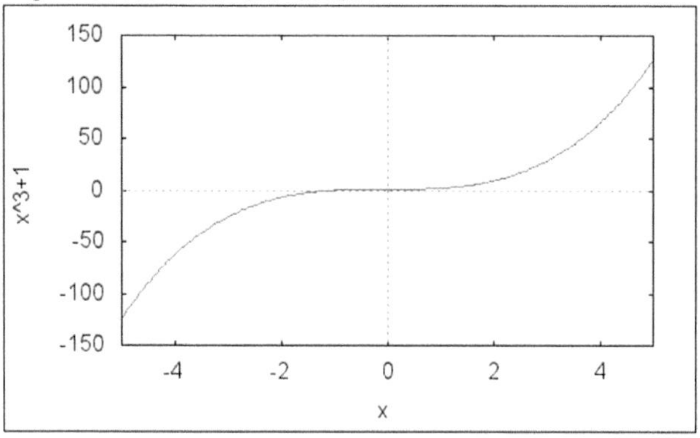

6) $p(x)=\dfrac{x^2-x}{x-1}$

```
(%i12) p(x):=(x^2-x)/(x-1);
```

(%o12) $p(x):=\dfrac{x^2-x}{x-1}$

```
(%i13) wxplot2d([p(x)], [x,-5,5])$
```

(%t13)

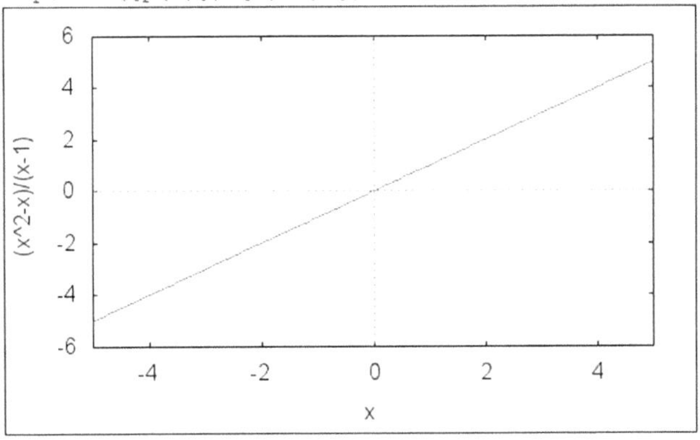

7) $k(x)=\sqrt{2x-1}$

```
(%i14) k(x):= sqrt(2*x-1);
```

(%o14) $k(x):=\sqrt{2\,x-1}$

```
(%i15) wxplot2d([k(x)], [x,-5,5])$
plot2d: expression evaluates to non-numeric value somewhere in plotting range.
```

(%t15)

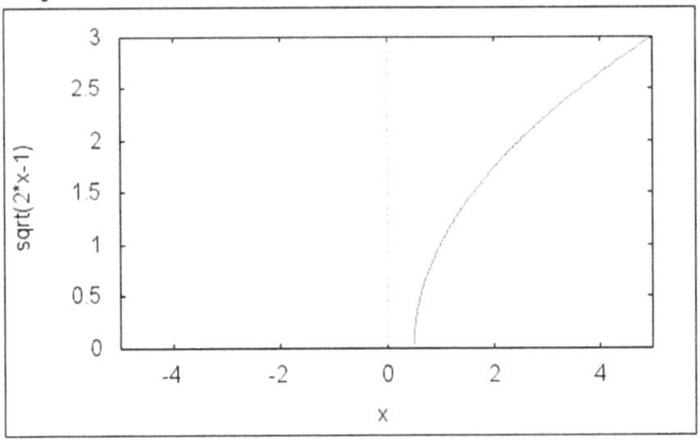

8) $f(p) = \sqrt{2-2p}$

```
(%i21) f(p):= sqrt(2-2*p);
(%o21) f(p):=√(2-2 p)

(%i23) wxplot2d([f(p)], [p,-5,5])$
plot2d: expression evaluates to non-numeric value somewhere in plotting range.
```

(%t23)

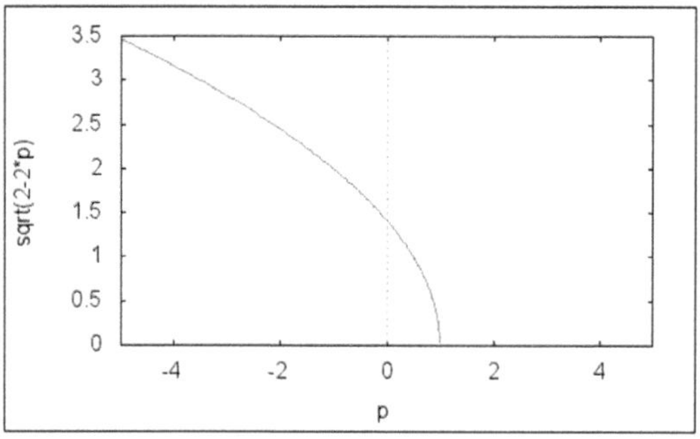

9) $F(x) = x^2 - 2x + 1$

```
(%i24) F(x):= x^2-2*x+1;
(%o24) F(x):=x^2 -2 x+1

(%i25) wxplot2d([F(x)], [x,-5,5])$
```

(%t25)

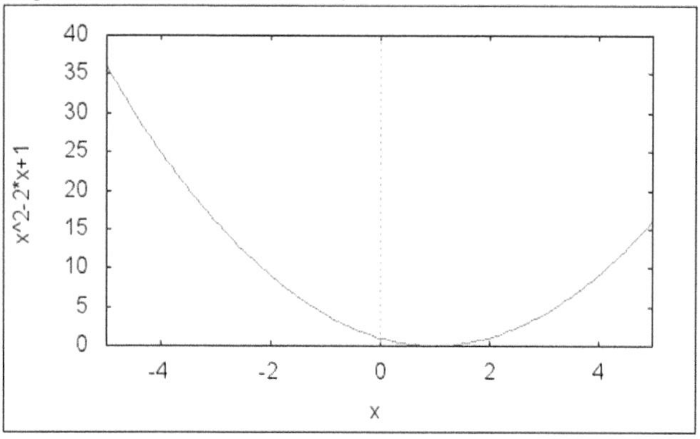

10) $f(t) = 2t + t^2$

```
(%i26) f(t):= 2*t+ t^2;
(%o26) f(t):=2 t+t^2
```

```
(%i27) wxplot2d([f(t)], [t,-5,5])$
```

(%t27)

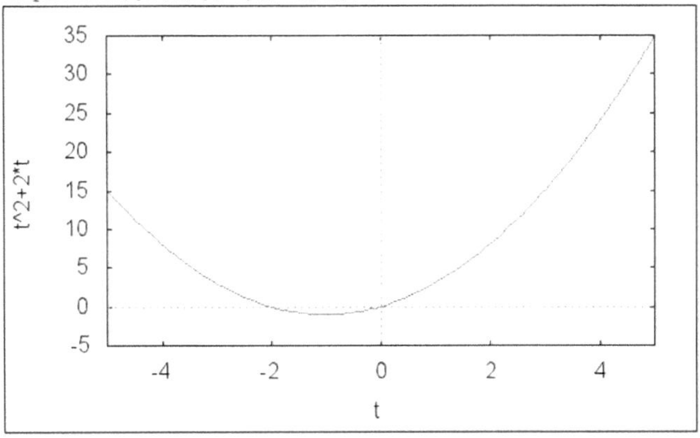

2. Draw each of the following functions and lines on the same graph using Maxima

1) $f(x) = x^2 - 2x + 1$ and $y = -2x$

Steps:

1) Go to maxima
2) Define the function
3) Go to Plot → Plot 2d
4) Write in the expression(s) box "$f(x)$,–2*x"
5) Click OK

Or

1) Go to maxima
2) Go to Plot → Plot 2d
3) Define the expressions of $f(x)$ and y
4) Click OK

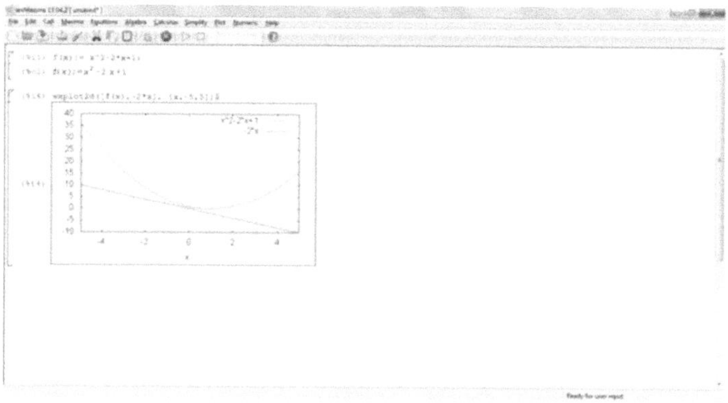

(See color insert.)

2) $f(x) = -x^2 + 5x$ and $y = xy = x$

```
(%i5) f(x) := -x^2+5*x;
(%o5) f(x):=-x^2+5 x

(%i6) wxplot2d([f(x),x], [x,-5,5])$
```

(%t6)

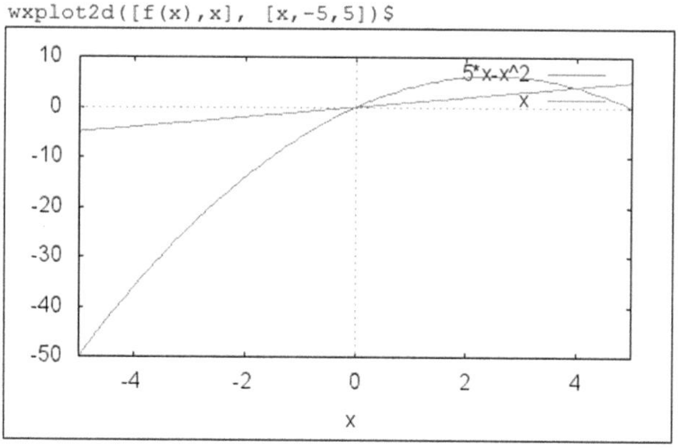

(See color insert.)

3) $f(x) = x^2 - 1$ and $y = -2x - 1$

```
(%i7) f(x):=x^2-1;
(%o7) f(x):=x^2-1
```

```
(%i8) wxplot2d([f(x),-2*x-1], [x,-5,5])$
```

(%t8)

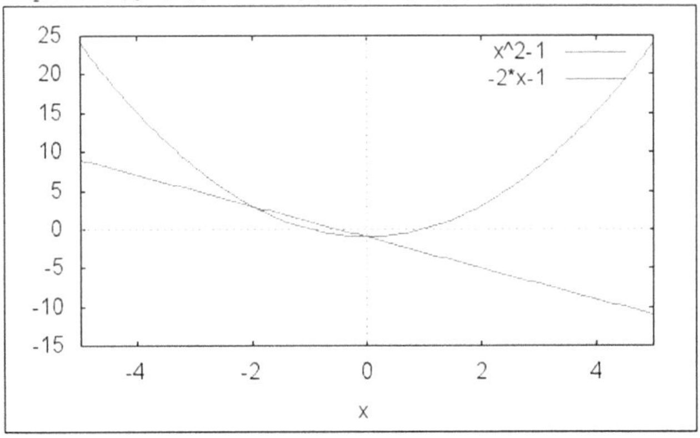

(See color insert.)

4) $f(x) = x^2 - 2$ and $y = 2x + 1$

```
(%i11) f(x):= x^2-2;
(%o11) f(x):=x^2-2
```

```
(%i12) wxplot2d([f(x),2*x+1], [x,-5,5])$
```

(%t12)

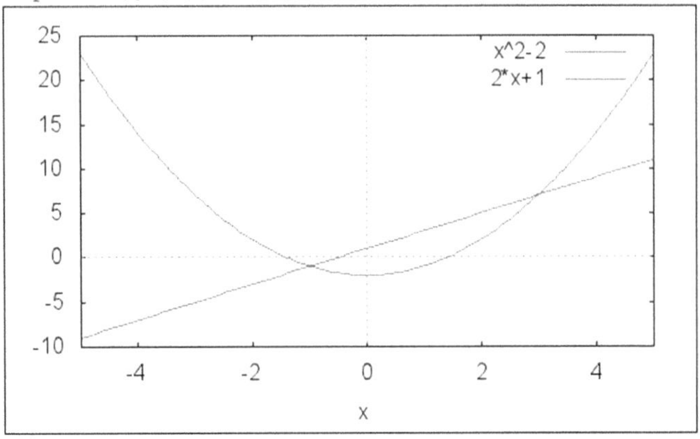

(See color insert.)

5) $f(x) = x^2 - 5x + 7$ and $y = x + 5$

```
(%i13) f(x):= x^2-5*x+7;
(%o13) f(x):=x^2 -5 x+7

(%i14) wxplot2d([f(x),x+5], [x,-5,5])$
```

(%t14)

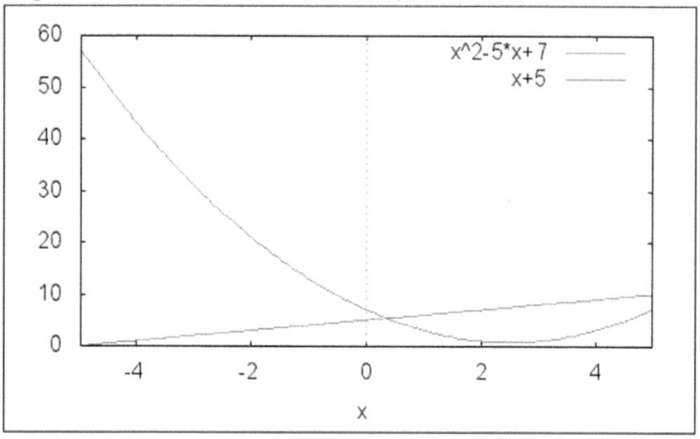

(See color insert.)

3. Given the two functions *f(x)* and *g(x)*.

$f(x) = 3 - x$ and $g(x) = x^2 - 1$

Find the following using Maxima:

a) $f(x) + g(x)$
b) $f(x) - g(x)$
c) $f(x) \times g(x)$
d) $f(x) \div g(x)$
e) $f(x) \circ g(x)$

Steps:

1) Go to Maxima
2) Define the two functions
a) Write in the command " f(x)+g(x)", then press Shift Enter
b) Write in the command " f(x)-g(x)", then press Shift Enter
c) Write in the command " f(x)*g(x)", then press Shift Enter
To write is in the simplest form go to Simplify → Simplify Expression
d) Write in the command " f(x)/g(x)", then press Shift Enter

To write is in the simplest form go to Simplify → Simplify Expression

e) Write in the command "f(g(x))", then press Shift Enter

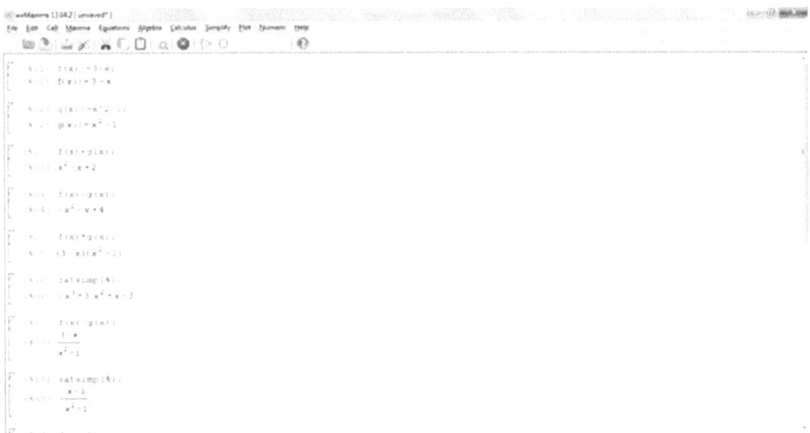

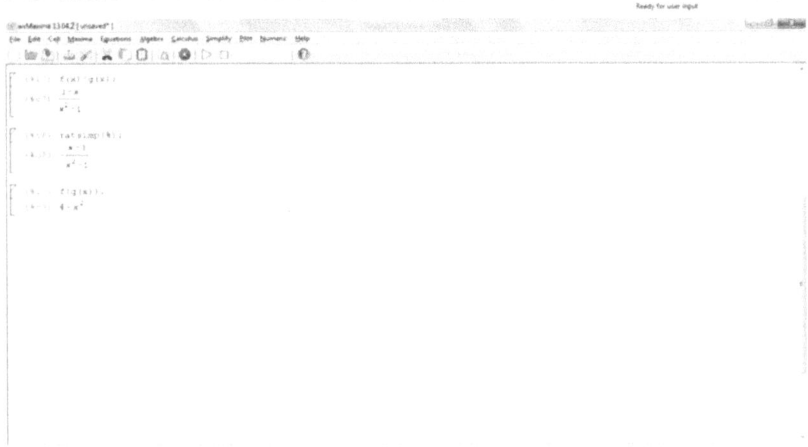

4. Given the two functions $f(x)$ and $g(x)$.

$f(x) = x^2 - 2x + 1$ and $g(x) = x^2 + 1$ and $g(x) = x^2 + 1$

Find the following using Maxima

a) $f(x) + g(x)$
b) $f(x) - g(x)$
c) $f(x) \times g(x)$

d) $f(x) \div g(x)$
e) $f(x) \circ g(x)$

```
(%i10) f(x):=x^2-2*x+1 ;
```
(%o10) $f(x):=x^2-2\,x+1$

```
(%i11) g(x):=x^2+1;
```
(%o11) $g(x):=x^2+1$

```
(%i12) f(x)+g(x);
```
(%o12) $2\,x^2-2\,x+2$

```
(%i13) f(x)-g(x);
```
(%o13) $-2\,x$

```
(%i14) f(x)*g(x);
```
(%o14) $(x^2+1)(x^2-2\,x+1)$

```
(%i15) ratsimp(%);
```
(%o15) $x^4-2\,x^3+2\,x^2-2\,x+1$

```
(%i16) f(x)/g(x);
```
(%o16) $\dfrac{x^2-2\,x+1}{x^2+1}$

```
(%i17) ratsimp(%);
```
(%o17) $\dfrac{x^2-2\,x+1}{x^2+1}$

```
(%i18) f(g(x));
```
(%o18) $(x^2+1)^2-2\,(x^2+1)+1$

APPLICATION

1. **Define and plot the following functions using Maxima**
   1) $H(t) = \dfrac{4 - t^2}{2 - t}$
   2) $g(x) = \sqrt{x - 5}$
   3) $y = \tan t - \cos t$
   4) $y = \dfrac{1}{1 + s}$
   5) $y = 3xy = 3x$
   6) $y = \dfrac{1}{x + 2}$
   7) $f(x) = \sqrt{3 - x}$
   8) $g(x) = \sqrt{x^2 - 1}$
   9) $g(x) = x^2 + 3x + 4$
   10) $h(x) = x^3 + 2$

2. **Draw each of the following functions and lines on the same graph using Maxima**
   1) $f(x) = x^2 - 2x + 1$ and $y = -2x + 2$
   2) $f(x) = -x^2 + 3$ and $y = x$
   3) $f(x) = x^2 - 4$ and $y = -2x + 1$
   4) $f(x) = x^2 - 9$ and $y = x + 3$
   5) $f(x) = x^2 - 6x + 5$ and $y = 2x + 5$

3. **Given the two functions *f(x)* and *g(x)*.**
   $f(x) = x - 2$ and $g(x) = x^2 + 3x + 4$

   Find the following using Maxima
   a) $f(x) + g(x)$
   b) $f(x) - g(x)$
   c) $f(x) \times g(x)$
   d) $f(x) \div g(x)$
   e) $f(x) \circ g(x)$

4. Given the two functions *f(x)* and *g(x)*.

$f(x) = x^3 + 2$ and $g(x) = x^2$

Find the following using Maxima

a) $f(x) + g(x)$
b) $f(x) - g(x)$
c) $f(x) \times g(x)$
d) $f(x) \div g(x)$
e) $f(x) \circ g(x)$

3.4 LIMITS AND CONTINUITY

Find the following limits using Maxima

1) $\lim_{x \to -2} \left(3x^4 + 2x^2 - x + 1\right)$

Steps:

1) Go to Maxima
2) Type in the command "**limit(3*x^4+2*x^2-x+1, x, –2)**"
3) Press Shift Enter

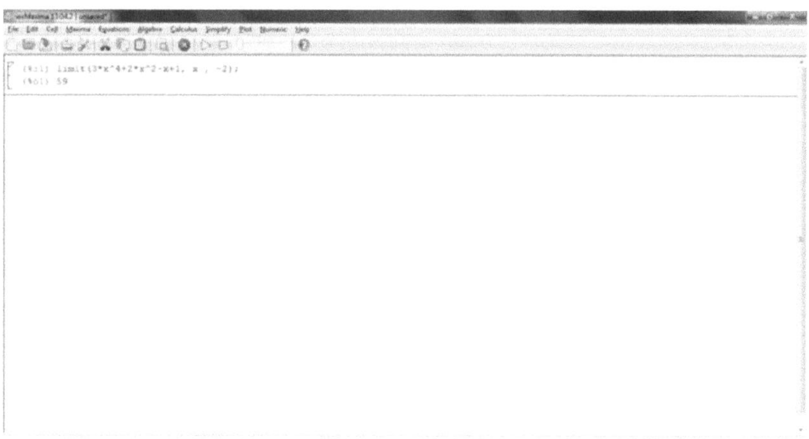

2) $\lim_{x \to -1} (x^4 - 3)(x^2 + 5x + 3)$

```
(%i3) limit((x^4-3*x)*(x^2+5*x+3), x , -1);
(%o3) -4
```

3) $\lim_{t \to -2} \frac{t^4 - 2}{2t^2 - 3t + 2}$

```
(%i4) limit((t^4-2)/(2*t^2-3*t+2), t , -2);
       7
(%o4)  -
       8
```

4) $\lim_{u \to -2} \sqrt{u^4 + 3u + 6}$

```
(%i5) limit(sqrt(u^4+3*u+6), u , -2);
(%o5) 4
```

5) $\lim_{t \to 2} \left(\frac{t^2 - 2}{t^3 - 3t + 5} \right)^2$

```
(%i6) limit(((t^2-2)/(t^3-3*t+5))^2, t , 2);
       4
(%o6)  --
       49
```

6) $\lim_{x \to 2} \sqrt{\frac{2x^2 + 1}{3x - 2}}$

```
(%i7) limit(sqrt((2*x^2+1)/(3*x-2)), x , 2);
       3
(%o7)  -
       2
```

7) $\lim_{x \to -4} \frac{x^2 + 5x + 4}{x^2 + 3x - 4}$

```
(%i8) limit((x^2+5*x+4)/(x^2+3*x-4), x , -4);
       3
(%o8)  -
       5
```

8) $\lim\limits_{x\to\infty} \dfrac{1}{2x+3}$

Steps:

1) Go to Maxima
2) Go to Calculus → Find Limit
3) Type the limit you want to find in the command box
4) Define your variable
5) Choose from special infinity
6) Click on OK

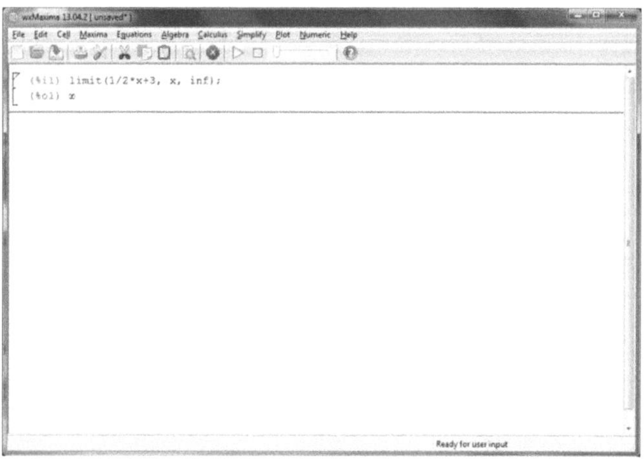

9) $\lim\limits_{x\to-\infty} \dfrac{1-x-x^3}{2x^2-7}$

```
(%i2) limit((1-x-x^3)/(2*x^2-7), x, minf);
(%o2) ∞
```

10) $\lim\limits_{x\to\infty} \dfrac{\left(2x^2+1\right)^2}{(x-1)^2\left(x^2+1\right)}$

```
(%i3) limit(((2*x^2+1)^2)/(((x-1)^2)*(x^2+1)), x, inf);
(%o3) 4
```

11) $\lim\limits_{x\to\infty}\dfrac{e^{3x}-e^{-3x}}{e^{3x}+e^{-3x}}$

```
(%i4) limit((%e^3*x-%e^-3*x)/(%e^3*x+%e^-3*x), x, inf);
        %e^6 - 1
(%o4)   --------
        %e^6 + 1
```

APPLICATION

Find the following limits using Maxima

1) $\lim\limits_{x\to 2}\dfrac{x^2-x+6}{x-2}$
2) $\lim\limits_{x\to -1}\dfrac{x^2-4x}{x^2-3x-4}$
3) $\lim\limits_{t\to -3}\dfrac{t^2-9}{2t^2+7t+3}$
4) $\lim\limits_{x\to -1}\dfrac{2x^2+3x+1}{x^2-2x-3}$
5) $\lim\limits_{t\to 1}\dfrac{t^4-1}{t^3-1}$
6) $\lim\limits_{x\to\infty}\sqrt{x^2+1}$
7) $\lim\limits_{x\to\infty}\dfrac{x^4-3x^2+x}{x^3-x+2}$
8) $\lim\limits_{x\to-\infty}\left(x^4+x^5\right)$
9) $\lim\limits_{x\to-\infty}\dfrac{1+x^6}{x^4+1}$
10) $\lim\limits_{x\to-\infty}x^2+1$

3.5 DERIVATIVES

Find the derivative of the following functions using Maxima

1) $f(t) = 2t^3 + t$

Steps:

1) Go to Maxima
2) Go to Calculus → Differentiate
3) Write the function in the command expression "2*t^3+t"
4) Here the variable is t to write t in the variable(s) box.

We need to differentiate one time so keep it 1, in case we need to find $f''(x)$ we just have to write 2 in the times box

5) Click OK

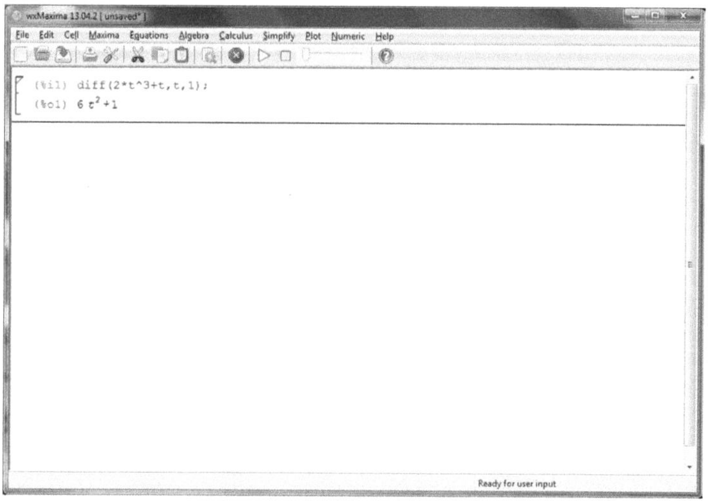

2) $f(t) = \dfrac{2t+1}{t+3}$

```
(%i2) diff((2*t+1)/(t+3),t,1);
```

$$\text{(\%o2)} \quad \frac{2}{t+3} - \frac{2\,t+1}{(t+3)^2}$$

```
(%i3) ratsimp(%);
```

$$\text{(\%o3)} \quad \frac{5}{t^2+6\,t+9}$$

3) $f(x) = x^{-2}$

```
(%i4) diff(x^-2,x,1);
```
(%o4) $-\frac{2}{x^3}$

4) $f(x) = \sqrt{1-2x}$

```
(%i5) diff(sqrt(1-2*x),x,1);
```
(%o5) $-\frac{1}{\sqrt{1-2\,x}}$

5) $f(x) = \frac{1}{2}x - \frac{1}{3}$

```
(%i7) diff(1/2*x-1/3,x,1);
```
(%o7) $\frac{1}{2}$

6) $f(t) = 5t - 9t^2$

```
(%i8) diff(5*t-9*t^2,t,1);
(%o8) 5-18 t
```

7) $g(x) = \sqrt{9-x}$

```
(%i9) diff(sqrt(9-x),x,1);
```
(%o9) $-\frac{1}{2\sqrt{9-x}}$

8) $f(x) = 186.5$

```
(%i12) diff(186.5,x,1);
(%o12) 0
```

9) $f(x) = 5x - 1$

```
(%i13) diff(5*x-1,x,1);
(%o13) 5
```

10) $F(x) = -4x^{10}$

```
(%i14) diff(-4*x^10,x,1);
(%o14) -40 x^9
```

11) $f(x) = x^3 - 4x + 6$

```
(%i15) diff(x^3-4*x+6,x,1);
(%o15) 3 x^2 -4
```

12) $f(t) = 1.4t^4 - 2.5t^2 + 6.7t^2 + 6.7$

```
(%i16) diff(1.4*t^4-2.5*t^2+6.7,t,1);
(%o16) 5.6 t^3 -5.0 t
```

APPLICATION

Find the derivative of the following functions using Maxima

1) $g(x) = x^2(1-2x)$
2) $h(x) = (x - 2)(2x + 3)$
3) $y = x^{-2/5}$
4) $B(y) = cy^{-6}$
5) $A(s) = -\frac{12}{s^5}$

6) $y = x^{5/3} - x^{2/3}$ $y = x^{5/3} - x^{2/3}$
7) $R(a) = (3a + 1)^2$
8) $h(t) = \sqrt[4]{t} - 4e^t$
9) $S(p) = \sqrt{p} - p$
10) $y = \sqrt{x}(x-1)$
11) $g(t) = t^3 \cos t$

3.6 INTEGRATION

Evaluate the integral using Maxima

1) $\int x^{\frac{4}{5}} dx$

Steps:

1) Go to Maxima
2) Click on Calculus →Integrate
3) Write your expression in the expression box
4) Click OK

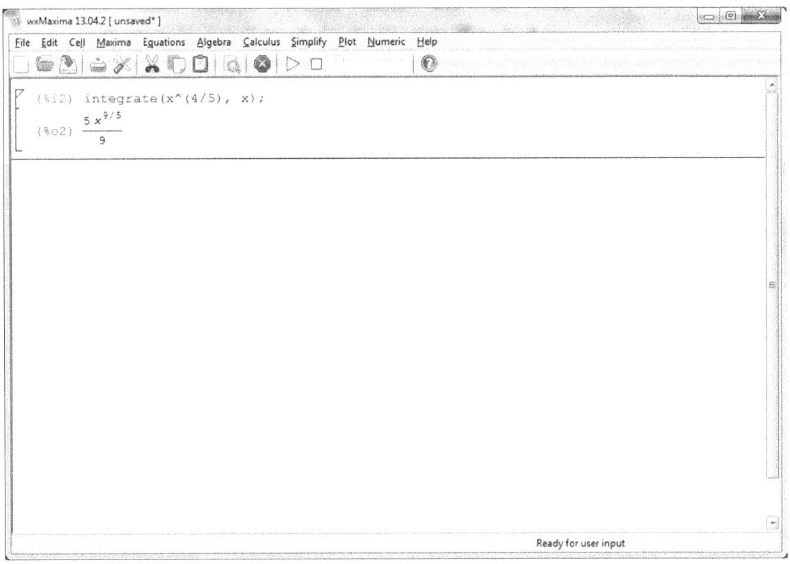

2) $\int_1^2 \frac{3}{t^4}\,dt$

```
(%i3) integrate(3/t^4, t, 1, 2);
```

(%o3) $\frac{7}{8}$

3) $\int_0^1 \left(3 + x\sqrt{x}\right)dx$

```
(%i4) integrate(3+x*sqrt(x), x, 0, 1);
```

(%o4) $\frac{17}{5}$

4) $\int_1^9 \frac{x-1}{\sqrt{x}}\,dx$

```
(%i5) integrate((x-1)/sqrt(x), x, 1, 9);
```

(%o5) $\frac{40}{3}$

5) $\int_1^2 (1+2y)^2\,dy$

```
(%i6) integrate((1+2*y)^2, y, 1, 2);
```

(%o6) $\frac{49}{3}$

6) $\int_1^2 \frac{4+u^2}{u^3}$

```
(%i7) integrate((4+u^2)/u^3, u, 1, 2);
```

(%o7) $\log(2)+\frac{3}{2}$

7) $\int_0^2 \left(6x^2-4x+5\right)dx$

```
(%i8) integrate(6*x^2-4*x+5, x, 0, 2);
(%o8) 18
```

8) $\int_1^3 (1+2x-4x^3)dx$

```
(%i9) integrate(1+2*x-4*x^3, x, 1, 3);
(%o9) -70
```

9) $\int_{-2}^0 (\frac{1}{2}t^4+\frac{1}{4}t^3-t)dt$

```
(%i10) integrate(1/2*t^4+1/4*t^3-t, t, 0, -2);
```

(%o10) $-\frac{21}{5}$

10) $\int_0^3 \left(1+6w^2-10w^4\right)dw$

```
(%i12) integrate(1+6*w^2-10*w^4, w, 0, 3);
(%o12) -429
```

Evaluate the integral using Maxima

1) $\int_{-1}^5 \left(1+3x\right)dx$

Steps:

1) Go to Maxima
2) Click on Calculus →Integrate
3) Write your expression in the expression box
4) Choose Definite integration, From –1 to 5

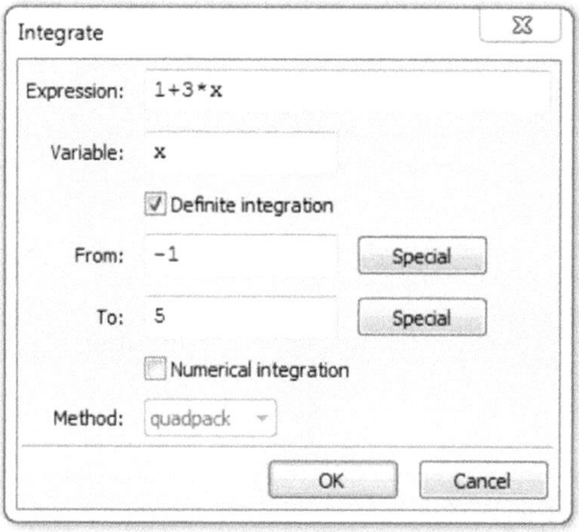

4) Click OK

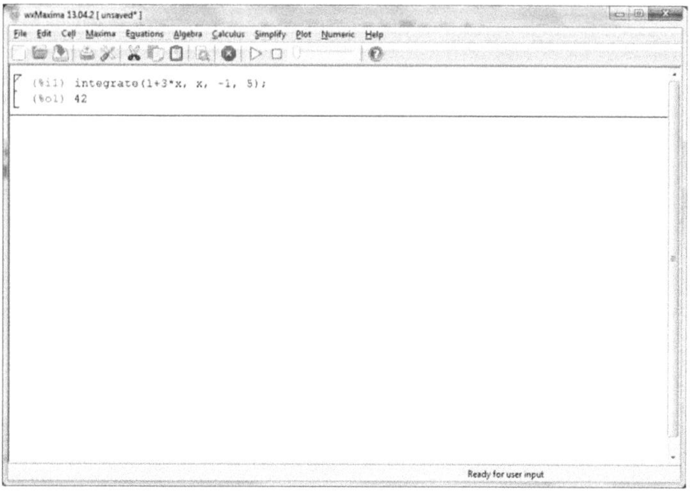

2) $\int_{-2}^{0}(x^2+x)dx$

```
(%i2) integrate(x^2+1, x, -2, 0);
      14
(%o2) --
      3
```

3) $\int_{1}^{4}\left(x^2+2x-5\right)dx$

```
(%i3) integrate(x^2+2*2-5, x, 1, 4);
(%o3) 18
```

4) $\int_{0}^{2}\left(2x-x^3\right)dx$

```
(%i4) integrate(2*x-x^3, x, 2, 0);
(%o4) 0
```

5) $\int_{0}^{1}\left(x^3-3x^2\right)dx$

```
(%i5) integrate(x^3-3*x^2, x, 0, 1);
        3
(%o5) - -
        4
```

6) $\int_{0}^{\pi}\sin 5x dx$

```
(%i8) integrate(sin (5*x), x, 0, %pi);
      2
(%o8) -
      5
```

7) $\int_{2}^{10} x^6 dx$

```
(%i9) integrate(x^6, x, 2, 10);
      9999872
(%o9) -------
         7
```

8) $\int_{-1}^{2} (1-x)\, dx$

```
(%i10) integrate(1-x, x, -1, 2);
       3
(%o10) -
       2
```

9) $\int_{0}^{9} \left(\frac{1}{3}x-2\right) dx$

```
(%i11) integrate(1/3*x-2, x, 0, 9);
         9
(%o11) - -
         2
```

10) $\int_{-1}^{2} |x|\, dx$

```
(%i12) integrate(abs(x), x, -1, 2);
       5
(%o12) -
       2
```

11) $\int_{\pi}^{\pi} \sin^2 x \cos^4 x\; dx$

```
(%i13) integrate(sin^2*(x)*cos^4*(x), x, %pi, %pi);
(%o13) 0
```

12) $\int_1^3 e^{x+2} dx$

```
(%i15) integrate((%e)^(x+2), x, 0, 1);
(%o15) %e^3 - %e^2
```

13) $\int_1^4 \sqrt{x} dx$

```
(%i16) integrate(sqrt(x), x, 1, 4);
(%o16) 14/3
```

AREA INTEGRALS USING MAXIMA

1) Plot $f(x) = e^{-0.2x}.cosx$ on [0; π/2] together with an area element at $x = 1$. Set up the integral A = ∫ dA and write dA in terms of x to obtain a definite integral. Finally, use Maxima to compute the area and make a shaded plot.

We start with the sketch of f(x) and dA:

```
f(x):=%e^(–0.2*x)*cos(x)$
wxdraw2d(
grid=true,
xaxis=true,
yaxis=true,
title="f(x) with dA near x=1",
color=black,
 explicit(f(x),x,0,%pi/2),
border=false,
color=red,
rectangle([1,0],[1.05,f(1.05)]),
color=black,
label(["dA",0.95,0.2])
);
```

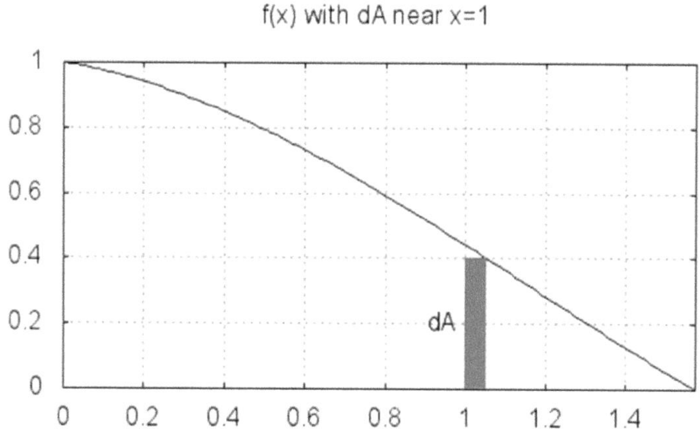

dA is just the product of height and width for the area element: f(x). dx. Now we set up the integral:

$$A = \int dA = \int_0^{\pi/2} f(x)\,dx = \int_0^{\pi/2} e^{-0.2x}.\cos x\,dx$$

Find the integral using Maxima, and we find a decimal approximation using float

```
(%i7) ratprint:false$

(%i8) integrate(f(x),x,0,%pi/2);
                 %pi
               - ---
                 10
        25 %e         5
(%o8)   ---------- + --
           26         26

(%i9) float(%);
(%o9) 0.89461797216216
```

We finish with a shaded plot:

```
wxdraw2d(
grid=true,
xrange=[–0.5,2],
yrange=[–0.5,1.2],
xaxis=true,
yaxis=true,
title=”Area Bounded by f(x) on [0,pi/2]”,
fill_color=grey,
filled_func=true,
filled_func=f(x),
 explicit(0,x,0,%pi/2),
filled_func=false,
color=black,
explicit(f(x),x,–0.5,2)
);
```

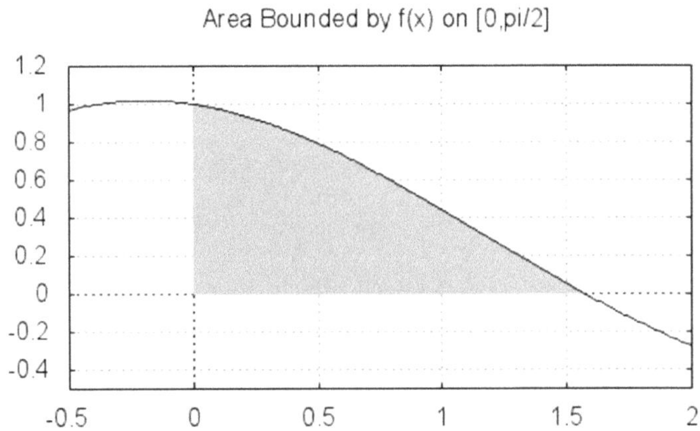

2) Compute the area bounded between $f(x) = 2 - x^2$ and $g(x) = x$

We start with a quick sketch of *f(x)* and *g(x)*

Steps:

1) Define the two functions
2) Go to Plot → Plot 2d

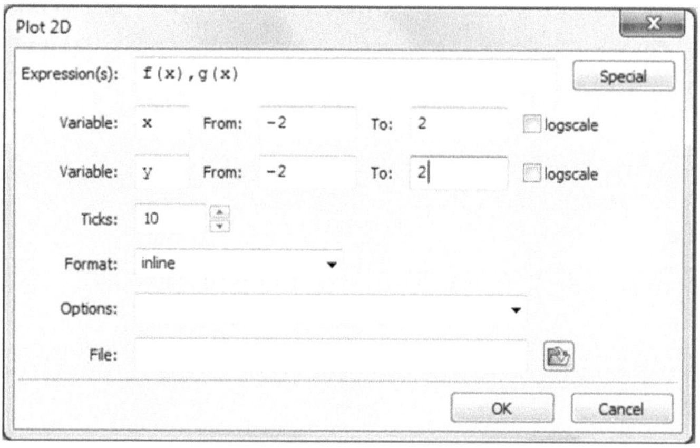

3) Click OK

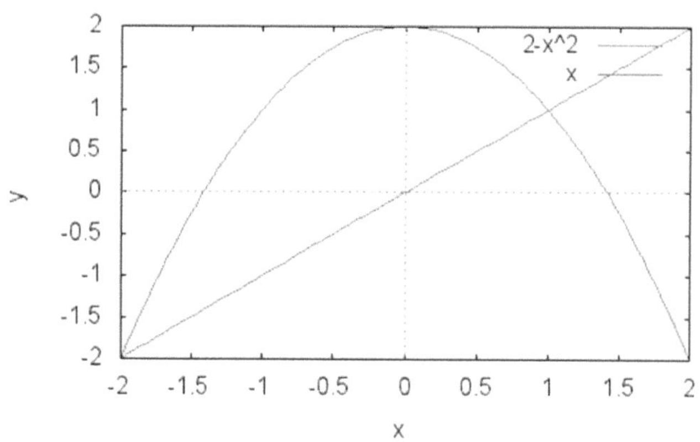

(See color insert.)

Now we determine the appropriate integration interval: the bounded area occurs between the two intersections of our curves.

Graphically it is for *x=1* and *x= –2*

Using Maxima:

```
(%i4) solve([f(x)=g(x)], [x]);
(%o4) [x=1,x=-2]
```

To find the area we just have to compute the integral

$$A = \int_{-2}^{1} \left[f(x) - g(x) \right] dx$$

```
(%i5) integrate(f(x)-g(x), x, -2, 1);
```

$$(\%o5)\ \frac{9}{2}$$

```
(%i6) float(%), numer;
(%o6) 4.5
```

We obtain A = 4.5 units.
This is the result graphically using Maxima
Write this code:

```
wxdraw2d(
grid=true,
xrange=[–2,2],
yrange=[–2,2],
xaxis=true,
yaxis=true,
title="Area Bounded between f(x) and g(x)",
fill_color=grey,
filled_func=true,
filled_func=f(x),
 explicit(g(x),x,–2,1),
filled_func=false,
color=black,
 explicit(f(x),x,–2,2),
color=red,
 explicit(g(x),x,–2,2)
);
```

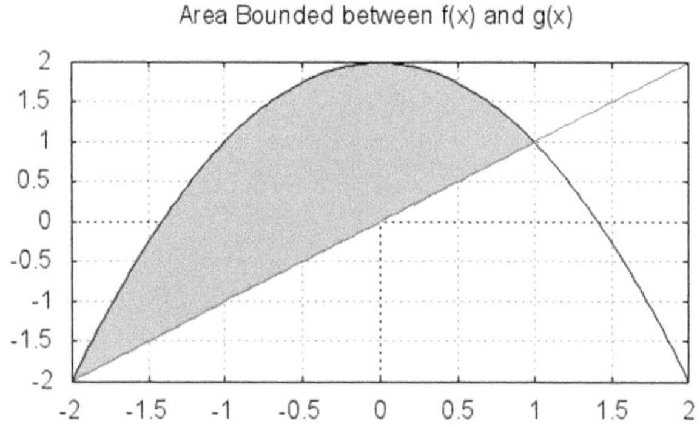

3) Compute the area bounded between $f(x)=x^3$ and $g(x)=6x$
Define the 2 functions:

```
(%i1) f(x):= x^3$
      g(x):=6*x$
```

Sketch them

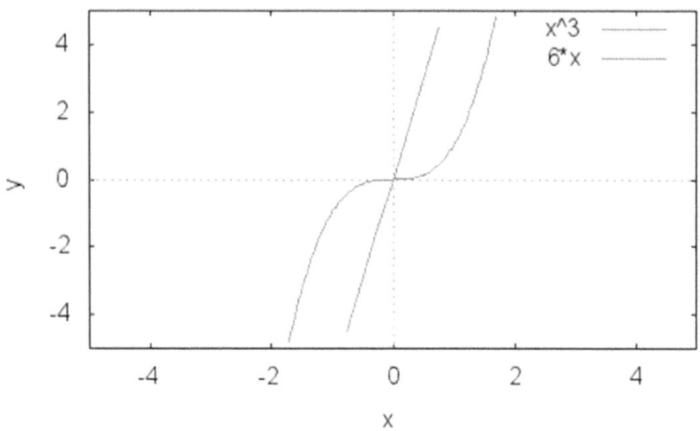

(See color insert.)

```
(%i7) solve([f(x)= g(x)], [x]);
```

(%o7) $[x=-\sqrt{6}, x=\sqrt{6}, x=0]$

```
(%i11) integrate(f(x)-g(x), x, -sqrt(6), sqrt(6));
(%o11) 0
```

Graphically:

wxdraw2d(
grid=true,
xrange=[–10,10],
yrange=[–10,10],
xaxis=true,
yaxis=true,
title="Area Bounded between f(x) and g(x)",
fill_color=grey,
filled_func=true,
filled_func=f(x),
explicit(g(x),x,-sqrt(6),sqrt(6)),
filled_func=false,
color=black,
explicit(f(x),x,–10,10),
color=red,
explicit(g(x),x,–10,10)
);

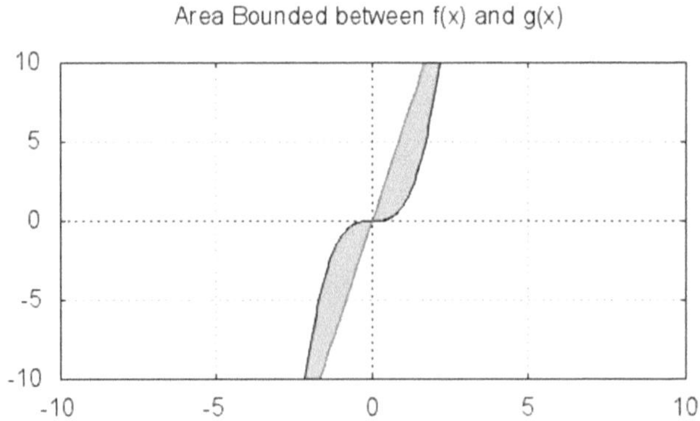

4) Compute the area bounded between $f(x) = x^2$ and $g(x) = \sqrt{x}$
   Define the 2 functions:

```
(%i1) f(x):=x^2$
      g(x)=sqrt(x)$
```

Sketch the functions:

```
wxplot2d([f(x),sqrt(x)], [x,-2,2], [y,-2,2])$
```

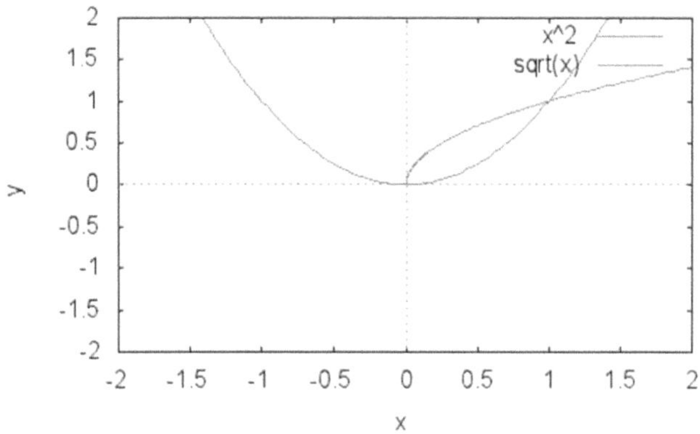

(See color insert.)

$x = 0$ and $x = 1$

```
(%i10) integrate(g(x)-f(x), x, 0, 1);
```

(%o10) $\frac{1}{3}$

Graphically:

wxdraw2d(
grid=true,
xrange=[–2,2],
yrange=[–2,2],
xaxis=true,
yaxis=true,
title="Area Bounded between f(x) and g(x)",
fill_color=grey,
filled_func=true,
filled_func=f(x),

explicit(g(x),x,0,1),
filled_func=false,
color=black,
explicit(f(x),x,–2,2),
color=red,
explicit(g(x),x,–2,2)
);

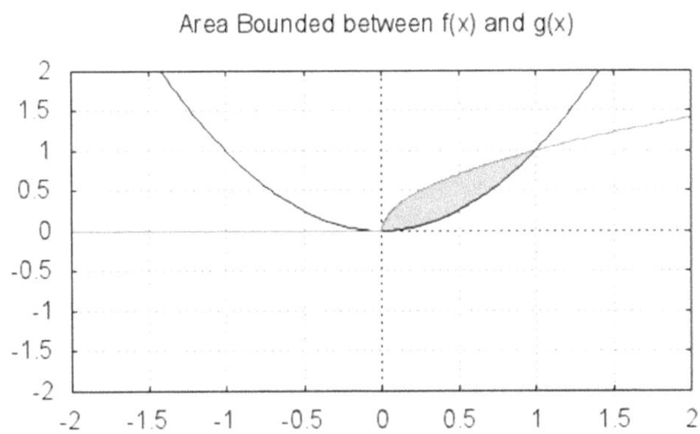

The Fundamental Theorem of Calculus

Use the fundamental Theorem of Calculus to compute $\int_{\pi/2}^{\pi} \sin x \, dx$

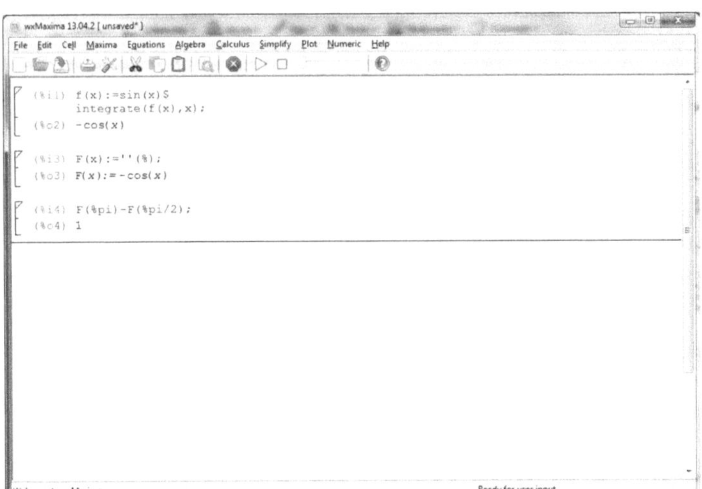

Use the fundamental Theorem of Calculus to compute $\int_1^3 \frac{1}{x^2}dx$

```
(%i5)  f(x):=1/x^2$
       integrate(f(x),x);
(%i6)
         1
(%o6)  - -
         x

(%i7)  F(x):=''(%);
                 1
(%o7)  F(x):= - -
                 x

(%i8)  F(3)-F(1);
       2
(%o8)  -
       3
```

INDEFINITE INTEGRAL

Find the general indefinite integral:

1) $\int (x^2 + x^{-2})dx$

```
(%i15)  integrate(x^2+x^(-2), x);
         3
        x    1
(%o15)  -- - -
        3    x
```

Or go to Calculus → Integration

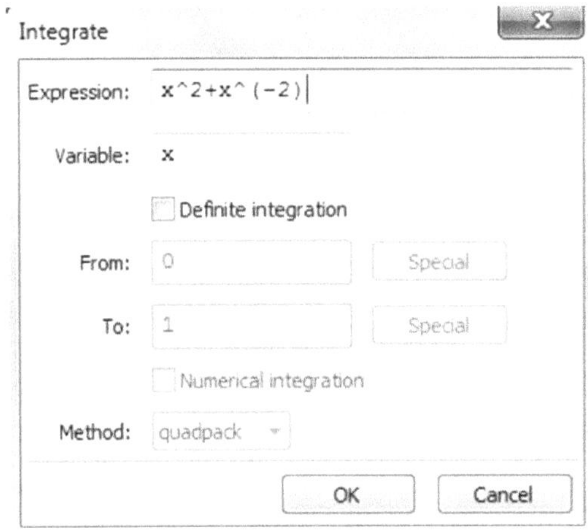

Click OK

2) $\int x^4 - \frac{1}{2}x^3 + \frac{1}{4}x - 2dx$

(%i19) integrate(x^4-1/2*x^3+1/4*x-2, x);

(%o19) $\frac{x^5}{5} - \frac{x^4}{8} + \frac{x^2}{8} - 2\,x$

3) $\int (u+4)(2u+1)\,du$

(%i20) integrate((u+4)*(2*u+1), u);

(%o20) $\frac{4\,u^3 + 27\,u^2 + 24\,u}{6}$

Evaluate the integral by making the given substitution

1) $\int 5x \sin x^2 dx \qquad u = x^2$

*(%i1) INTEGRAND:(5*x)*sin(x^2)*del(x)$*

(%i2) solve(diff(u)=diff(x^2),del(x));

*(%o2) [del(x)=del(u)/(2*x)]*

(%i3) %[1];

*(%o3) del(x)=del(u)/(2*x)*

(%i4) subst(rhs(%),del(x),INTEGRAND);

*(%o4) (5*sin(x^2)*del(u))/2*

(%i5) subst(u,x^2,%);

*(%o5) (5*sin(u)*del(u))/2*

(%i6) integrate(coeff(%,del(u)),u);

*(%o6) -(5*cos(u))/2*

(%i7) subst(x^2,u,%);

*(%o7) -(5*cos(x^2))/2(%i6) integrate(coeff(%,del(u)),u);*

*(%o6) -(5*cos(u))/2*

(%i7) subst(x^2,u,%);

*(%o7) -(5*cos(x^2))/2*

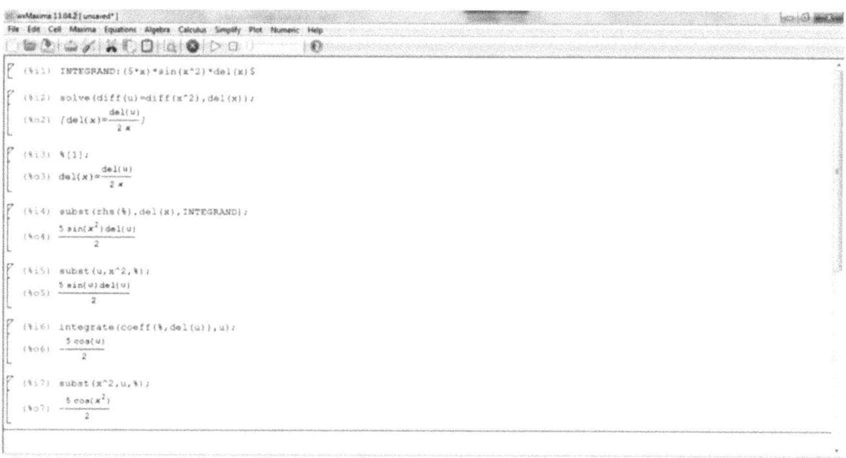

2) $\int \cos 3x \, dx \qquad u = 3x$

```
(%i8) INTEGRAND:cos(3*x)*del(x)$

(%i10) solve(diff(u)=diff(3*x),del(x));
```

(%o10) $[\mathrm{del}(x)=\frac{\mathrm{del}(u)}{3}]$

```
(%i11) %[1];
```

(%o11) $\mathrm{del}(x)=\frac{\mathrm{del}(u)}{3}$

```
(%i12) subst(rhs(%),del(x),INTEGRAND);
```

(%o12) $\frac{\cos(3\,x)\,\mathrm{del}(u)}{3}$

```
(%i13) subst(u,3*x,%);
```

(%o13) $\frac{\cos(u)\,\mathrm{del}(u)}{3}$

```
(%i14) integrate(coeff(%,del(u)),u);
```

(%o14) $\frac{\sin(u)}{3}$

```
(%i15) subst(3*x,u,%);
```

(%o15) $\frac{\sin(3\,x)}{3}$

3) $\int \cos^3 x \; sinx \; dx \qquad u = cos\, x$

```
(%i27) INTEGRAND:(cos(x))^3*sin(x)*del(x)$

(%i28) solve(diff(u)=diff(cos(x)),del(x));
                 del(u)
(%o28) [del(x)=- ------]
                 sin(x)

(%i29) %[1];
                del(u)
(%o29) del(x)=- ------
                sin(x)

(%i30) subst(rhs(%),del(x),INTEGRAND);
              3
(%o30) -cos(x)  del(u)

(%i32) subst(u,cos(x),%);
         3
(%o32) -u  del(u)

(%i33) integrate(coeff(%,del(u)),u);
         4
        u
(%o33) --
        4

(%i34) subst(cos(x),u,%);
             4
       cos(x)
(%o34) -------
          4
```

INTEGRATION BY PARTS

Evaluate the integral using integration by parts using Maxima.

1) $\int x.\sin x \; dx$

Steps: You have to choose u and dv, then find v by computing $\int dv$

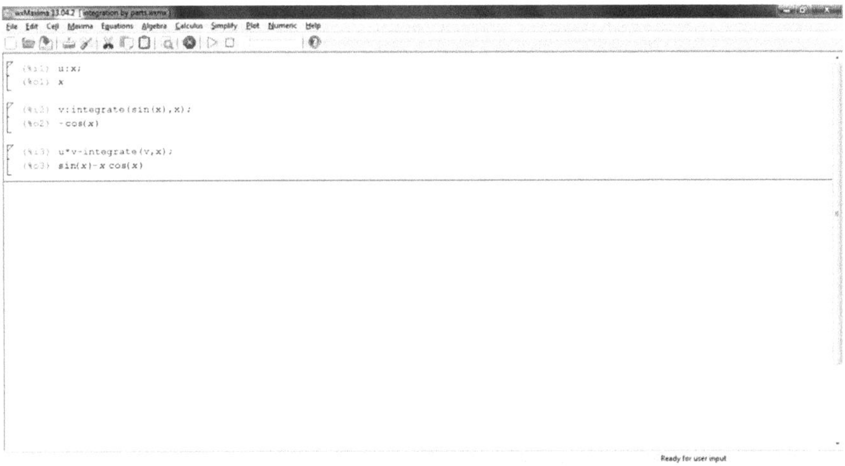

2) $\int x \cos x \, dx$

```
(%i16) u:x;
(%o16) x

(%i17) v:integrate(cos(x),x);
(%o17) sin(x)

(%i18) u*v-integrate(v,x);
(%o18) x sin(x)+cos(x)
```

3) $\int x^2 sinx\ dx$

```
(%i1) u:x^2;
(%o1) x^2

(%i2) diff(u);
(%o2) 2 x del(x)

(%i3) v:integrate(sin(x),x);
(%o3) -cos(x)

(%i4) Term1:u*v;
(%o4) -x^2 cos(x)

(%i5) Integrand1:v*diff(u);
(%o5) -2 x cos(x) del(x)
```

The first parts iteration is complete, yielding $-x^2.\cos x - \int -2x.\cos x dx$

```
(%i6) u1:-2*x$
      diff(u1);
(%o7) -2 del(x)

(%i8) v1:integrate(cos(x),x);
(%o8) sin(x)

(%i9) Term2:u1*v1;
(%o9) -2 x sin(x)

(%i10) Integrand2:v1*diff(u1);
(%o10) -2 sin(x) del(x)
```

The second parts integration yields $-2x.\sin x - \int -2sinx\ dx$.
Finally, we put together the final result: $u.v - (u1v1 - \int v1du1)$

```
(%i11) Term1-(Term2-integrate(coeff(Integrand2,del(x)),x));
(%o11) 2 x sin(x)-x^2 cos(x)+2 cos(x)
```

INTEGRATION OF RATIONAL FUNCTIONS BY PARTIAL FRACTIONS

Write out the form of the partial fraction decomposition of the function.

1) $\dfrac{x-4}{3x^3+5x^2+4x+2}$

We start by factoring the denominator and proposing the partial fractions decomposition as an equation:

```
(%i1) R:(x-4)/(3*x^3+5*x^2+4*x+2)$
      factor(denom(R));
(%o2) (x+1)(3 x^2+2 x+2)
```

```
(%i3) EQN:R=(A/(x+1))+(B*x+C)/(3*x^2+2*x+2);
```

$$(\%o3)\quad \frac{x-4}{3x^3+5x^2+4x+2}=\frac{C+xB}{3x^2+2x+2}+\frac{A}{x+1}$$

Now we multiply both sides of the equation by the original denominator, expand the result to a polynomial and produce a list of equations by comparing the coefficients of each power of x on the left and right sides:

```
(%i4) EQN*(denom(R));
```

$$(\%o4)\quad x-4=(3x^3+5x^2+4x+2)\left(\frac{C+xB}{3x^2+2x+2}+\frac{A}{x+1}\right)$$

```
(%i5) ratsimp(%);
```

$$(\%o5)\quad x-4=(x+1)C+(x^2+x)B+(3x^2+2x+2)A$$

```
(%i6) expand(%);
```

$$(\%o6)\quad x-4=xC+C+x^2B+xB+3x^2A+2xA+2A$$

```
(%i7) EQN1:coeff(lhs(%o6),x,2)=coeff(rhs(%o6),x,2);
      EQN2:coeff(lhs(%o6),x,1)=coeff(rhs(%o6),x,1);
      EQN3:coeff(lhs(%o6),x,0)=coeff(rhs(%o6),x,0);
```

$$(\%o7)\quad 0=B+3A$$

(%i8)

$$(\%o8)\quad 1=C+B+2A$$

(%i9)

$$(\%o9)\quad -4=C+2A$$

We solve this system of equations for A, B, and C and substitute into the original decomposition:

```
(%i10) solve([EQN1,EQN2,EQN3],[A,B,C]);
```

$$(\%o10)\quad [[A=-\frac{5}{3},B=5,C=-\frac{2}{3}]]$$

```
(%i11) sublis([A=-5/3,B=5,C=-2/3],rhs(EQN));
```

$$(\%o11)\quad \frac{5x-\frac{2}{3}}{3x^2+2x+2}-\frac{5}{3(x+1)}$$

2) $\dfrac{x^3+x}{x-1}$

Since the degree of the numerator is greater than the degree of the numerator, we perform the long division

Steps:

1) Go to Maxima
2) Define p:x^3+x$
   q: x–1$
   divide(p,q);

```
(%i4) p:x^3+x$
      q:x-1$
      divide(p,q);
(%o6) [x^2+x+2,2]
```

So $\dfrac{x^3+x}{x-1} = x^2+x+2+\dfrac{2}{x-1}$

3) $\dfrac{x^4-2x^2+4x+11}{x^3-x^2-x+1}$

The first step is to divide.

```
(%i7) p:x^4-2*x^2+4*x+1$
      q:x^3-x^2-x+1$
      divide(p,q);
(%o9) [x+1,4 x]
```

So $\dfrac{x^4-2x^2+4x+11}{x^3-x^2-x+1} = x+1+\dfrac{4x}{x^3-x^2-x+1}$

We have to find the partial fraction decomposition of $\dfrac{4x}{x^3-x^2-x+1}$

```
R:4*x/(x^3-x^2-x+1)$
factor(denom(R));
(x-1)^2 (x+1)
```

So $\dfrac{4x}{x^3-x^2-x+1} = \dfrac{A}{x-1}+\dfrac{B}{(x-1)^2}+\dfrac{C}{x+1}$

```
(%i18) EQN:R=(A/(x-1)+B/(x-1)^2+C/(x+1));
```

(%o18) $$\frac{4\,x}{x^3-x^2-x+1}=\frac{C}{x+1}+\frac{B}{(x-1)^2}+\frac{A}{x-1}$$

Now we multiply both sides of the equation by the original denominator, expand the result to a polynomial and produce a list of equations by comparing the coefficients of each power of x on the left and right sides:

```
(%i19) EQN*(denom(R));
```

(%o19) $$4\,x=(x^3-x^2-x+1)\left(\frac{C}{x+1}+\frac{B}{(x-1)^2}+\frac{A}{x-1}\right)$$

```
(%i20) ratsimp(%);
```

(%o20) $$4\,x=(x^2-2\,x+1)\,C+(x+1)\,B+(x^2-1)\,A$$

```
(%i21) expand(%);
```

(%o21) $$4\,x=x^2\,C-2\,x\,C+C+x\,B+B+x^2\,A-A$$

```
(%i22) E1:coeff(lhs(%o21),x,2)=coeff(rhs(%o21),x,2);
```

(%o22) $$0=C+A$$

```
(%i23) E2:coeff(lhs(%o21),x,1)=coeff(rhs(%o21),x,1);
```

(%o23) $$4=B-2\,C$$

```
(%i24) E3:coeff(lhs(%o21),x,0)=coeff(rhs(%o21),x,0);
```

(%o24) $$0=C+B-A$$

We solve this system of equations for A, B, and C and substitute into the original decomposition:

```
(%i25) solve([E1,E2,E3],[A,B,C]);
(%o25) [[A=1,B=2,C=-1]]
```

```
(%i26) sublis([A=1,B=2,C=-1],rhs(EQN));
```

$$(\%o26)\quad -\frac{1}{x+1}+\frac{1}{x-1}+\frac{2}{(x-1)^2}$$

$$\frac{x^4-2x^2+4x+11}{x^3-x^2-x+1}=x+1+\frac{1}{x-1}+\frac{2}{(x-1)^2}-\frac{1}{x+1}$$

IMPROPER INTEGRATE

Determine whether the integral is convergent or divergent using Maxima

1) $\int_1^{\infty}\frac{1}{x^2}dx$

```
(%i1) integrate(1/x^2, x, 1, inf);
(%o1) 1
```

2) $\int_{-\infty}^{0}\frac{1}{3-4x}dx$

```
(%i2) integrate(1/(3-4*x), x, minf, 0);
defint: integral is divergent.
 -- an error. To debug this try: debugmode(true);
```

3) $\int_1^4\frac{1}{(x-3)^2}$

```
(%i1) integrate(1/(x-3)^2, x, 1, 4);
defint: integral is divergent.
 -- an error. To debug this try: debugmode(true);
```

APPLICATION

1. Evaluate the integral using Maxima

1) $\int_{0}^{2}(2x-3)(4x^2+1)dx$

2) $\int_{-1}^{1}t(1-t^2)dt$

3) $\int_{0}^{3}xdx$

4) $\int_{1}^{2}x^2dx$

5) $\int_{0}^{2}(3x^2+4x)dx$

6) $\int_{2}^{6}4dx$

7) $\int_{3}^{4}(x^2+3x-2)dx$

8) $\int_{\pi/6}^{2\pi/3}\sin xdx$

9) $\int_{0}^{\pi/2}\cos xdx$

10) $\int_{3}^{5}(4x^3+3x+2)dx$

11) $\int_{-3}^{0}\left(1+\sqrt{9-x^2}\right)dx$

12) $\int_{-5}^{5}\left(x-\sqrt{25-x^2}\right)dx$

13) $\int_{0}^{10}|x-5|dx$

14) $\int_{0}^{2} e^{x+5} dx$

15) $\int_{\frac{\pi}{4}}^{\frac{\pi}{3}} \tan x dx$

16) $\int_{0}^{2} \frac{1}{1+x^2} dx$

17) $\int_{0}^{2} xe^{-x} dx$

2. Plot $f(x) = x^2 - 4$ on [0; 22] together with an area element at x = 1. Set up the integral $A = \int dA$ and write dA in terms of x to obtain a definite integral. Finally, use Maxima to compute the area and make a shaded plot.
3. Compute the area bounded between $f(x) = 3x^3 - x^2 - 10x$ and $g(x) = -x^2 + 2x$
4. Compute the area bounded between $f(x) = \sqrt{x}$ and $g(x) = 2x^2$
5. Use the fundamental Theorem of Calculus to compute $\int_{1}^{3} \frac{1}{x^2+4} dx$
6. Use the fundamental Theorem of Calculus to compute $\int_{0}^{3} (x^3 - 6x) dx$
7. Find the general indefinite integral:

   a) $\int v(v^2 + v)^2 dv$

   b) $\int (\sin(\sin x + \sinh x) dx$

   c) $\int \frac{t^2 - 1}{t^4 - 1} dt$

8. Evaluate the integral by making the given substitution

   a) $\int x\left(4 + x^2\right)^{10} dx \qquad u = 4 + x^2$

   b) $\int \frac{dt}{(1-6t)^4} \qquad u = 1 - 6t$

   c) $\int \frac{\sec^2\left(\frac{1}{x}\right)}{x^2} dx \qquad u = 1/x$

9. Evaluate the integral using integration by parts using Maxima.
   a) $\int x.cos5x\ dx$
   b) $\int \arctan 4t\ dt$
   c) $\int re^{r/2}dr$
10. Write out the form of the partial fraction decomposition of the function
   a) $10/5x^2 - 2x^2$
   b) $\dfrac{1+6x}{(4x-3)(2x+5)}$
   c) $\dfrac{x^2-1}{x^3+x^2+x}$
   d) $\dfrac{x^2+2x-1}{2x^3+3x^2-2x}$
11. Determine whether the integral is convergent or divergent using Maxima
   a) $\int_0^\infty \frac{x}{x^3+1}dx$

3.7 SEQUENCES AND SERIES

Find the first 20 partial sums for the series $\sum_{n=1}^{\infty} 1/3^n$

```
(%i1) a(n):=1/3^n;
```

$$(\%o1)\quad a(n):=\frac{1}{3^n}$$

```
(%i2) (print("n ...... partial sum"),
      print("1 ......",a(1)),
      for i:2 thru 20 do
      (S:float(sum(a(n),n,1,i)),
      print(i, "......",S))
      );
```

```
n ...... partial sum
          1
1 ...... -
          3
2 ...... 0.44444444444444
3 ...... 0.48148148148148
4 ...... 0.49382716049383
5 ...... 0.49794238683128
6 ...... 0.49931412894376
7 ...... 0.49977137631459
8 ...... 0.49992379210486
9 ...... 0.49997459736829
10 ...... 0.4999915324561
11 ...... 0.49999717748537
12 ...... 0.49999905916179
13 ...... 0.49999968638726
14 ...... 0.49999989546242
15 ...... 0.49999996515414
16 ...... 0.49999998838471
17 ...... 0.49999999612824
18 ...... 0.49999999870941
19 ...... 0.4999999995698
20 ...... 0.4999999998566
 (%o2) done
```

Or Go to Calculus→ Calculate sum
Define your expression

```
(%i6) sum(1/(3^n), n, 1, 20), simpsum;
        1743392200
(%o6) ----------
        3486784401

(%i7) float(%);
(%o7) 0.499999998566
```

Calculate the first eight terms of the sequence of partial sums

1) $\sum_{n=1}^{\infty} \frac{1}{n^3}$

```
(%i8) sum(1/(n^3), n, 1, 8), simpsum;
        78708473
(%o8) --------
        65856000

(%i9) float(%), numer;
(%o9) 1.19516024356171
```

2) $\sum_{n=1}^{\infty} \cos n$

```
(%i11) sum(cos(n), n, 1, 8), simpsum;
(%o11) cos(8)+cos(7)+cos(6)+cos(5)+cos(4)+cos(3)+cos(2)+cos(1)

(%i12) float(%), numer;
(%o12) 0.33275404450522
```

3) $\sum_{n=1}^{\infty} \frac{12}{(-5)^n}$

```
(%i13) sum(12/(-5)^n, n, 1, 8), simpsum;
        781248
(%o13) - ------
        390625

(%i14) float(%), numer;
(%o14) -1.99999488
```

INTEGRAL TEST

Test the following series for convergence or divergence

1) $\sum_{n=1}^{\infty} \frac{1}{n^2+1}$

The function $\frac{1}{x^2+1}$ is continuous, positive, and decreasing on $[1,\infty)$ so we use the integral test:

$$\sum_{1}^{\infty} \frac{1}{x^2+1}dx$$

```
(%i1) integrate(1/(x^2+1), x, 1, inf);
       π
(%o1)  -
       4
```

Convergent

2) $\sum_{n=1}^{\infty} \frac{1}{n}$

.

```
(%i1) f(x):=1/x$
      integrate(f(x),x,1,inf);
defint: integral is divergent.
 -- an error. To debug this try: debugmode(true);
```

THE RATIO AND ROOT TESTS

1) Apply both the ratio and root tests to show that $\sum_{n=1}^{\infty} \frac{2^n}{e^{0.8n-3}}$ converges absolutely.

```
(%i1) a(n):=2^n/%e^(0.8*n-3)$
      limit(a(n+1)/a(n),n,inf);
(%o2) 0.89865792823444

(%i3) float(limit((a(n))^(1/n),n,inf));
rat: replaced -0.8 by -4/5 = -0.8
(%o3) 0.89865792823444
```

In each case, the limit yields constant less than 1, so the series converges absolutely.

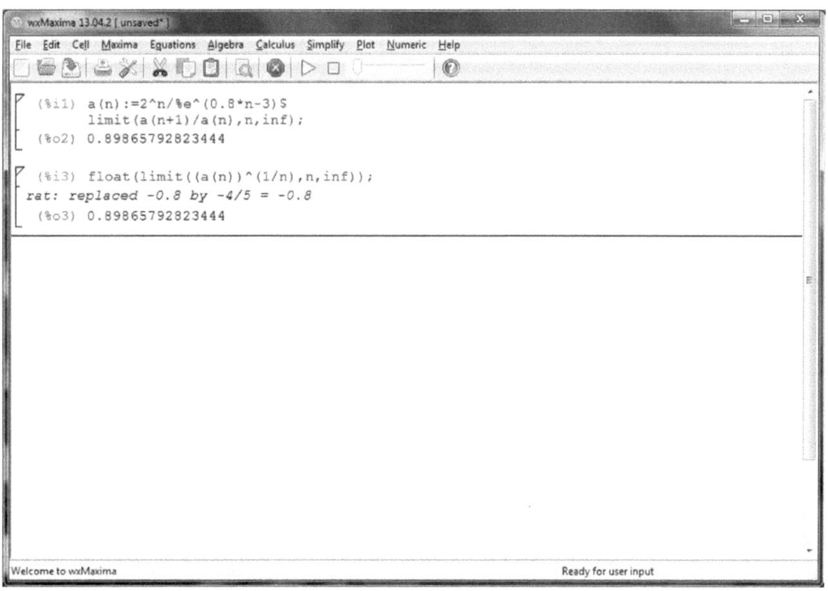

2) Test the series $\sum_{n=1}^{\infty} \frac{n^{\ln 2n}}{n!}$ for convergence.

```
(%i4) a(n):=n^(log(2*n))/n!$
      limit(a(n+1)/a(n),n,inf);
(%o5) 0

(%i6) float(limit((a(n))^(1/n),n,inf));
(%o6) 0.0
```

In each case, we obtain a constant less than 1, so the series is absolutely convergent.

POWER SERIES

RADIUS AND INTERVAL OF CONVERGENCE

1) Find the values of x for which the series $\sum_{n=0}^{\infty}\frac{x^n}{n!}$ converges.

```
(%i5) A(n):=x^n/n!$
      limit(abs(A(n+1)/A(n)),n,inf);
(%o6) 0
```

The result is less than 1 regardless of the value of x. The radius of convergence is infinite, and the interval of convergence is $(-\infty, \infty)$

2) Find the values of x for which the series $\sum_{n=0}^{\infty}\frac{(x-2)^n}{n(n+1)}$ converges.

```
(%i7) A(n):=(x-2)^n/(n*(n+1))$
      limit(abs(A(n+1)/A(n)),n,inf);
(%o8) |x-2|
```

$|x-2| < 1| < 1 \rightarrow\ -1 < x-2 < 1 \rightarrow 1 < x < 3$, so the series converges on (1, 3).

Testing the endpoint x = 1, we obtain the altering series $\sum_{n=0}^{\infty}\frac{(-1)^n}{n(n+1)}$ which converges because $\lim_{n\to\infty}\frac{1}{n(n+1}=0$ and $\frac{1}{(n+1)(n+2)}<\frac{1}{n(n+1)}$

Testing the endpoint x =3, we obtain $\sum_{n=0}^{\infty}\frac{1}{n(n+1)}$ which converges to $\sum_{n=0}^{\infty}\frac{1}{n^2}$ because $\frac{1}{n(n+1)}<\frac{1}{n^2}$. Thus, the interval of convergence is [1, 3].

TAYLOR SERIES

Drive the first five Taylor coefficients by defining $f(x) = a_0 + a_1(x - c)^2 + \ldots f(x) = a_0 + a_1(x - c) + a^2 (x - c)^2 + \ldots$ and computing $f(c), f'(c), f''(c)$, and so on.

f(x):=a_0+a_1(x-c)+a_2*(x-c)^2+a_3*(x-c)^3+a_4*(x-c)^4+a_5*(x-c)^5$*
(for n:0 thru 4 do
(N:subst([x=c],diff(f(x),x,n)),
print("f^(",n,")(x)=",diff(f(x),x,n)),
print("f^(",n,")(c)=",N))
);

```
f(x):=a_0+a_1*(x-c)+a_2*(x-c)^2+a_3*(x-c)^3+a_4*(x-c)^4+a_5*(x-c)^5$
(for n:0 thru 4 do
 (N:subst([x=c],diff(f(x),x,n)),
print("f^(",n,")(x)=",diff(f(x),x,n)),
print("f^(",n,")(c)=",N))
);
```

```
f^(0)(x)=a_5(x-c)^5+a_4(x-c)^4+a_3(x-c)^3+a_2(x-c)^2+a_1(x-c)+a_0
f^(0)(c)=a_0
f^(1)(x)=5 a_5(x-c)^4+4 a_4(x-c)^3+3 a_3(x-c)^2+2 a_2(x-c)+a_1
f^(1)(c)=a_1
f^(2)(x)=20 a_5(x-c)^3+12 a_4(x-c)^2+6 a_3(x-c)+2 a_2
f^(2)(c)=2 a_2
f^(3)(x)=60 a_5(x-c)^2+24 a_4(x-c)+6 a_3
f^(3)(c)=6 a_3
f^(4)(x)=120 a_5(x-c)+24 a_4
f^(4)(c)=24 a_4
 (%o6) done
```

APPLICATION

1) Calculate the first ten terms of the sequence of partial sums

   a) $\sum_{n=1}^{\infty} \frac{n}{\sqrt{n^2+4}}$

   b) $\sum_{n=1}^{\infty} \frac{7^{n+1}}{10^n}$

   c) $\sum_{n=2}^{\infty} \frac{1}{n(n+2)}$

2) Test the following series for convergence or divergence

   a) $\sum_{n=1}^{\infty} \frac{1}{n^3}$ (Integral test)

   b) $\sum_{n=1}^{\infty} \frac{n}{n^2+1}$ (Integral test)

   c) $\sum_{n=1}^{\infty} (-1)^n \frac{n^3}{3^n}$ (Ratio test)

   d) $\sum_{n=1}^{\infty} \frac{n^n}{n!}$ (Ratio test)

3) Find the values of x for which the series $\sum_{n=0}^{\infty} n!x^n$ converges.

4) Find the values of x for which the series $\sum_{n=0}^{\infty} \frac{(x-3)^n}{n}$ converges.

3.8 CURVES IN 3-D SPACE

1) Sketch the curve whose vector equation is:

$r(t) = cos(t)\, i + sin(t)\, j + tk$

Steps:

1) Go to Maxima
2) Define "r(t):=[cos(t),sin(t),t]"
3) Press Shift Enter
4) Load(draw)
5) Press Shift Enter
6) draw3d(parametric(cos(t), sin(t), t, t, –4, 4));

```
(%i1) r(t):=[cos(t),sin(t),t];
(%o1) r(t):=[cos(t),sin(t),t]

(%i2) load(draw);
Loading C:/Users/pauly.awad/maxima/binary/binary-gcl/share/draw/grcommon.o
Finished loading C:/Users/pauly.awad/maxima/binary/binary-gcl/share/draw/grcommon.o
Loading C:/Users/pauly.awad/maxima/binary/binary-gcl/share/draw/gnuplot.o
Finished loading C:/Users/pauly.awad/maxima/binary/binary-gcl/share/draw/gnuplot.o
Loading C:/Users/pauly.awad/maxima/binary/binary-gcl/share/draw/vtk.o
Finished loading C:/Users/pauly.awad/maxima/binary/binary-gcl/share/draw/vtk.o
Loading C:/Users/pauly.awad/maxima/binary/binary-gcl/share/draw/picture.o
Finished loading C:/Users/pauly.awad/maxima/binary/binary-gcl/share/draw/picture.o
(%o2) C:/PROGRA~1/MAXIMA~1.2/share/maxima/5.31.2/share/draw/draw.lisp

(%i3) draw3d(parametric(cos(t), sin(t), t, t, -4, 4));
```

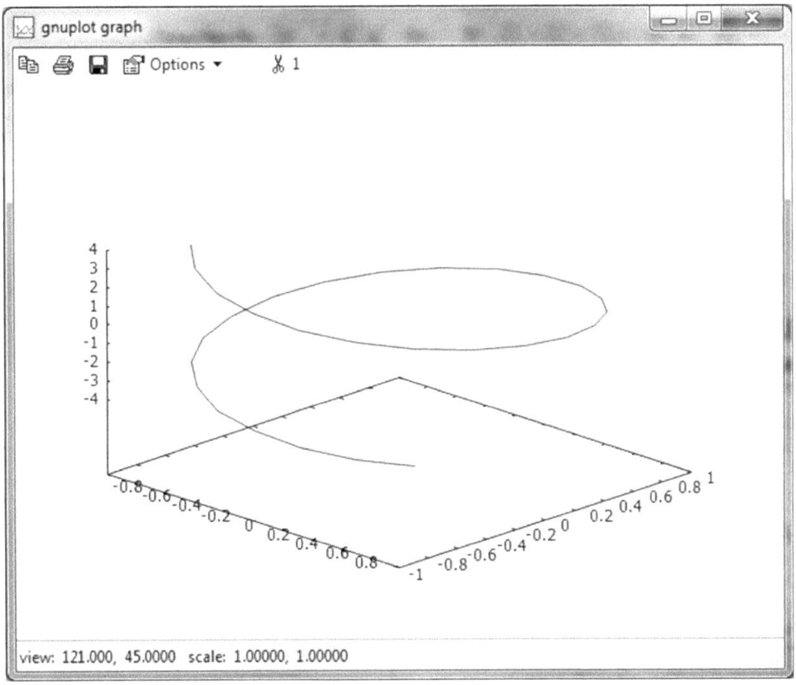

b) Find $\lim_{t \to 2} r(t)$

```
(%i2) limit(r(t),t,0);
(%o2) [1,0,0]
```

c) Find the derivative

```
(%i3) diff(r(t),t);
(%o3) [-sin(t),cos(t),1]
```

d) Find the integral

```
(%i4) integrate(r(t),t);
```

(%o4) $[\sin(t), -\cos(t), \frac{t^2}{2}]$

e) Find the unit tangent vector T as the normalized derivative of r

```
(%i5) define(rp(t), diff(r(t),t));
(%o5) rp(t):=[-sin(t),cos(t),1]

(%i6) load(eigen);
(%o6) C:/PROGRA~1/MAXIMA~1.2/share/maxima/5.31.2/share/matrix/eigen.mac

(%i7) uvect(rp(t));
```

(%o7) $[-\frac{\sin(t)}{\sqrt{\sin(t)^2+\cos(t)^2+1}}, \frac{\cos(t)}{\sqrt{\sin(t)^2+\cos(t)^2+1}}, \frac{1}{\sqrt{\sin(t)^2+\cos(t)^2+1}}]$

```
(%i8) trigsimp(%);
```

(%o8) $[-\frac{\sin(t)}{\sqrt{2}}, \frac{\cos(t)}{\sqrt{2}}, \frac{1}{\sqrt{2}}]$

2) a) Sketch the curve whose vector equation is:

$r(t) = ti + cos(t)j + sin(t)k$

```
(%i1) r(t):=[t,cos(t),sin(t)];
(%o1) r(t):=[t,cos(t),sin(t)]

(%i2) load(draw);
Loading C:/Users/pauly.awad/maxima/binary/binary-gcl/share/draw/grcommon.o
Finished loading C:/Users/pauly.awad/maxima/binary/binary-gcl/share/draw/grcommon.o
Loading C:/Users/pauly.awad/maxima/binary/binary-gcl/share/draw/gnuplot.o
Finished loading C:/Users/pauly.awad/maxima/binary/binary-gcl/share/draw/gnuplot.o
Loading C:/Users/pauly.awad/maxima/binary/binary-gcl/share/draw/vtk.o
Finished loading C:/Users/pauly.awad/maxima/binary/binary-gcl/share/draw/vtk.o
Loading C:/Users/pauly.awad/maxima/binary/binary-gcl/share/draw/picture.o
Finished loading C:/Users/pauly.awad/maxima/binary/binary-gcl/share/draw/picture.o
(%o2) C:/PROGRA~1/MAXIMA~1.2/share/maxima/5.31.2/share/draw/draw.lisp

(%i3) draw3d(parametric(t, cos(t), sin(t), t, -6, 6));
```

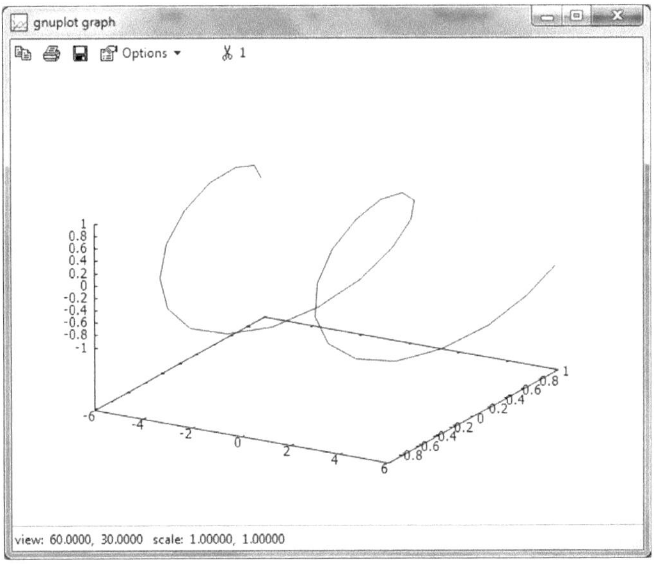

b) Find $\lim_{t \to 2} r(t)$

```
(%i4) limit(r(t),t,2);
(%o4) [2 ,cos(2),sin(2)]
```

c) Find the derivative

```
(%i5) diff(r(t),t);
(%o5) [1 ,-sin(t),cos(t)]
```

d) Find the integral

```
(%i6) integrate(r(t),t);
```

(%o6) $[\frac{t^2}{2}, \sin(t), -\cos(t)]$

e) Find the unit tangent vector T as the normalized derivative of r.

```
(%i7) define(rp(t), diff(r(t),t));
(%o7) rp(t):=[1,-sin(t),cos(t)]

(%i8) load(eigen);
(%o8) C:/PROGRA~1/MAXIMA~1.2/share/maxima/5.31.2/share/matrix/eigen.mac

(%i9) uvect(rp(t));
```

(%o9) $[\frac{1}{\sqrt{\sin(t)^2+\cos(t)^2+1}},-\frac{\sin(t)}{\sqrt{\sin(t)^2+\cos(t)^2+1}},\frac{\cos(t)}{\sqrt{\sin(t)^2+\cos(t)^2+1}}]$

```
(%i10) trigsimp(%);
```

(%o10) $[\frac{1}{\sqrt{2}},-\frac{\sin(t)}{\sqrt{2}},\frac{\cos(t)}{\sqrt{2}}]$

```
(%i11) define(T(t),%);
```

(%o11) $T(t):=[\frac{1}{\sqrt{2}},-\frac{\sin(t)}{\sqrt{2}},\frac{\cos(t)}{\sqrt{2}}]$

ARC LENGTH AND CURVATURE

There is no special Maxima function for the curvature but we can do it with the formulas we learned.

r(t) = ti + cos(t)j + sin(t)k

```
(%i1) r(t):=[t,cos(t),sint];
(%o1) r(t):=[t,cos(t),sint]

(%i2) rp(t):=[1,-sin(t),cos(t)];
(%o2) rp(t):=[1,-sin(t),cos(t)]

(%i3) Tp(t):=[0,-cos(t),sin(t)]/sqrt(2);
```

(%o3) $Tp(t):=\frac{[0,-\cos(t),\sin(t)]}{\sqrt{2}}$

```
(%i5) sqrt(Tp(t).Tp(t))/sqrt(rp(t).rp(t));
```

(%o5) $\frac{\sqrt{\frac{\sin(t)^2}{2}+\frac{\cos(t)^2}{2}}}{\sqrt{\sin(t)^2+\cos(t)^2+1}}$

```
(%i6) trigsimp(%);
```

(%o6) $\frac{1}{2}$

APPLICATION

1) a) Sketch the curve whose vector equation is:

$$r(t) = (1+t^3)i + te^{-t}j + sin(2t)k$$

b) Find $\lim_{t\to 2} r(t)$

c) Find the derivative

d) Find the integral

e) Find the unit tangent vector T as the normalized derivative of r.

2) Find the curvature of

a) $r(t) = t^3j + t^2k$

b) $r(t) = 3ti + 4\ sin(t)j + 4\ cos\ (t)k$

3.9 FUNCTIONS OF SEVERAL VARIABLES

1) Sketch the graph of:

a) $f(x, y) = (x^2 - y^2)^2)$

Steps:

1) Go to Maxima
2) Define the function
3) Load(draw)
4) Press Shift Enter
5) draw3d(explicit(f(x,y),x,–4,4,y,–4,4)); (to get a plain surface plot)

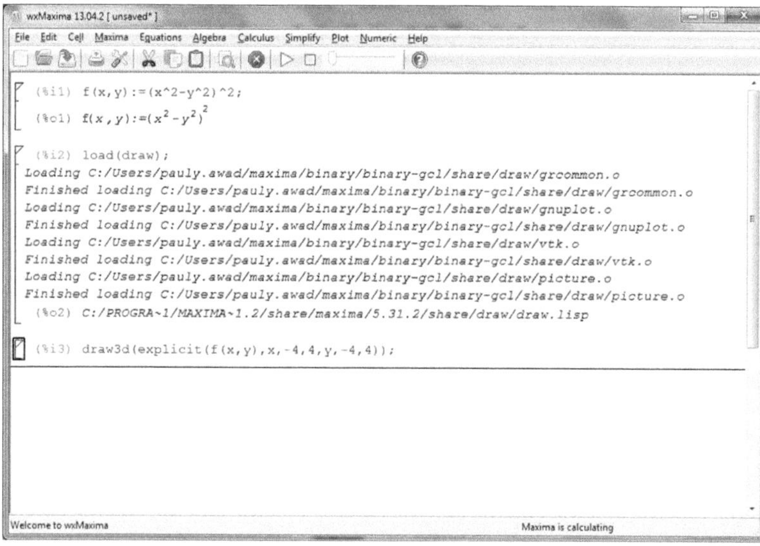

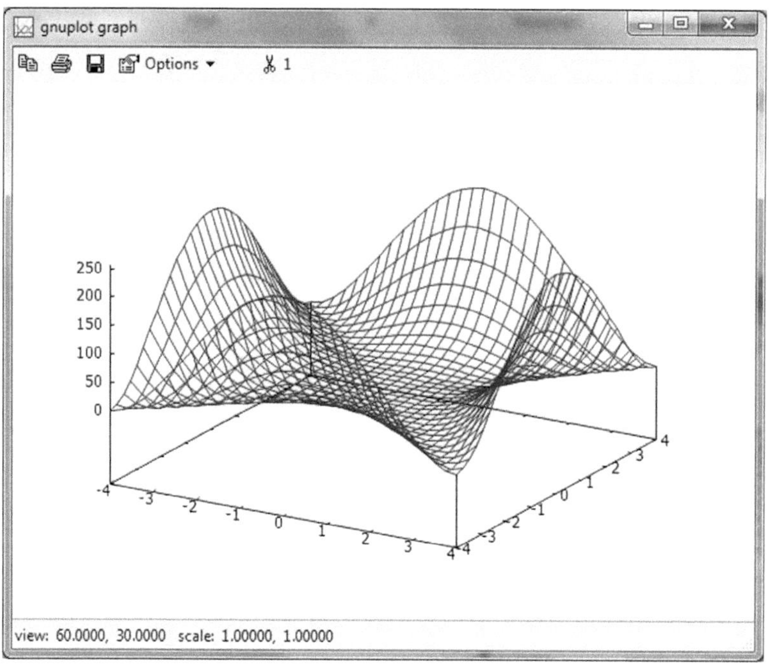

6) draw 3d(enhanced 3d = true, explicit (f(x,y),x,–4,4,y,–4,4)); (an enhanced surface plot)

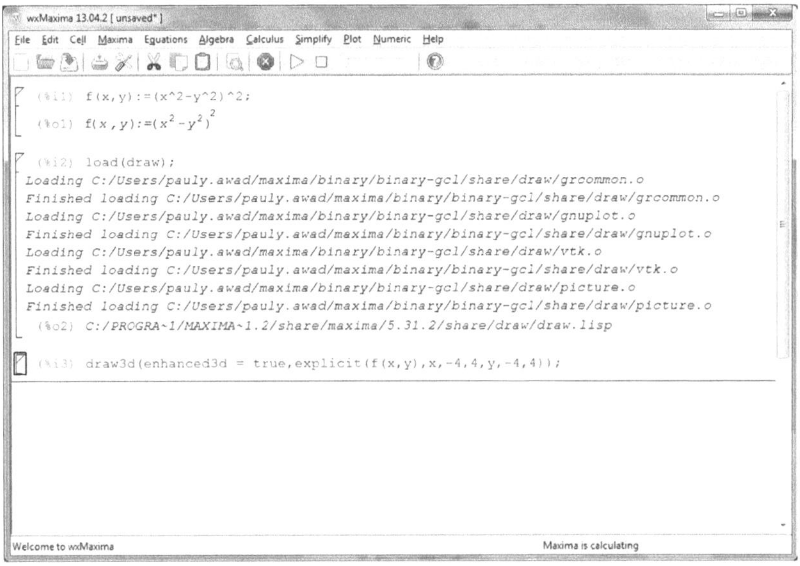

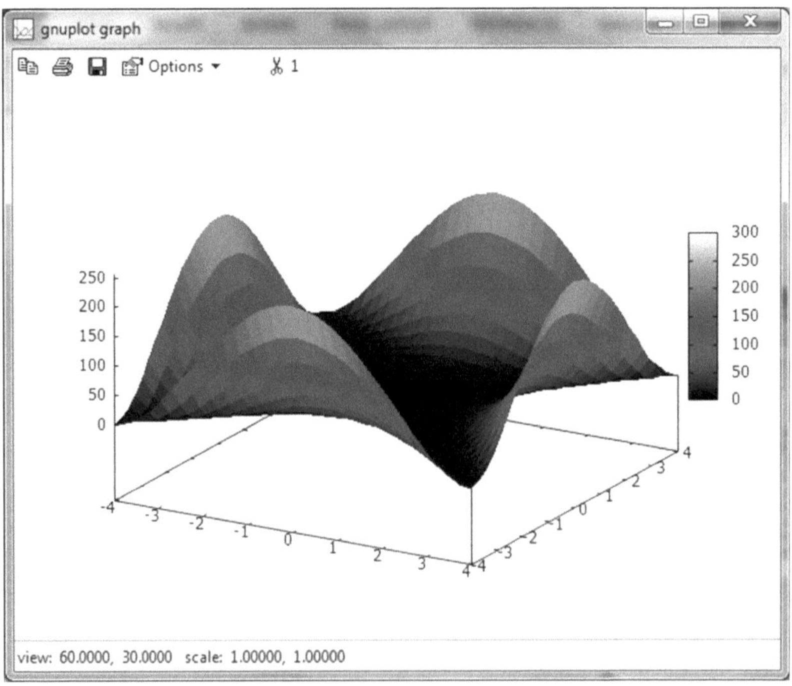

(See color insert.)

b) $g(x,y)=\sqrt{9-x^2-y^2}$

```
(%i1) g(x,y):= sqrt(9-x^2-y^2);
```

(%o1) $g(x,y):=\sqrt{9-x^2-y^2}$

```
(%i2) load(draw);
Loading C:/Users/pauly.awad/maxima/binary/binary-gcl/share/draw/grcommon.o
Finished loading C:/Users/pauly.awad/maxima/binary/binary-gcl/share/draw/grcommon.o
Loading C:/Users/pauly.awad/maxima/binary/binary-gcl/share/draw/gnuplot.o
Finished loading C:/Users/pauly.awad/maxima/binary/binary-gcl/share/draw/gnuplot.o
Loading C:/Users/pauly.awad/maxima/binary/binary-gcl/share/draw/vtk.o
Finished loading C:/Users/pauly.awad/maxima/binary/binary-gcl/share/draw/vtk.o
Loading C:/Users/pauly.awad/maxima/binary/binary-gcl/share/draw/picture.o
Finished loading C:/Users/pauly.awad/maxima/binary/binary-gcl/share/draw/picture.o
(%o2) C:/PROGRA~1/MAXIMA~1.2/share/maxima/5.31.2/share/draw/draw.lisp

(%i3) draw3d(explicit(g(x,y),x,-4,4,y,-4,4));
```

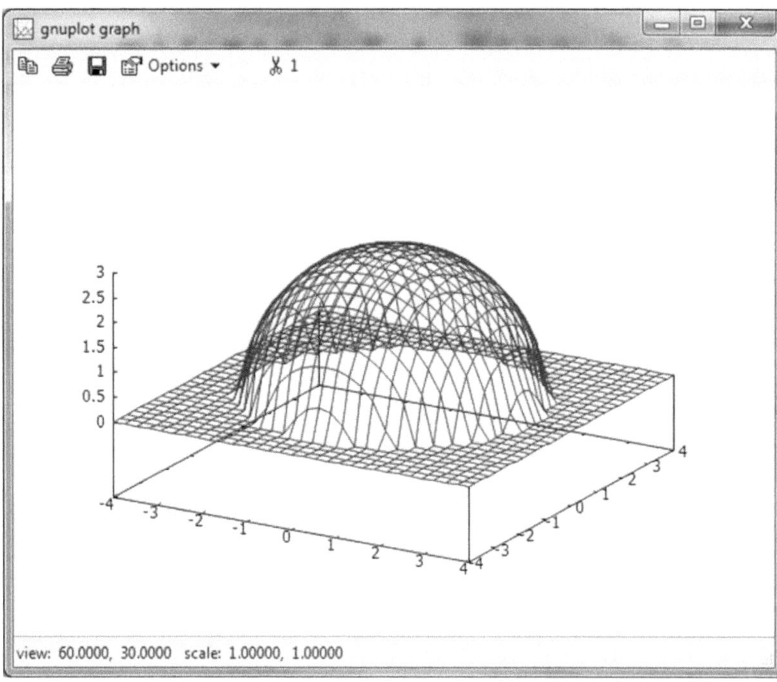

```
(%i1) g(x,y):= sqrt(9-x^2-y^2);
(%o1) g(x,y):=√(9-x²-y²)

(%i2) load(draw);
Loading C:/Users/pauly.awad/maxima/binary/binary-gcl/share/draw/grcommon.o
Finished loading C:/Users/pauly.awad/maxima/binary/binary-gcl/share/draw/grcommon.o
Loading C:/Users/pauly.awad/maxima/binary/binary-gcl/share/draw/gnuplot.o
Finished loading C:/Users/pauly.awad/maxima/binary/binary-gcl/share/draw/gnuplot.o
Loading C:/Users/pauly.awad/maxima/binary/binary-gcl/share/draw/vtk.o
Finished loading C:/Users/pauly.awad/maxima/binary/binary-gcl/share/draw/vtk.o
Loading C:/Users/pauly.awad/maxima/binary/binary-gcl/share/draw/picture.o
Finished loading C:/Users/pauly.awad/maxima/binary/binary-gcl/share/draw/picture.o
(%o2) C:/PROGRA~1/MAXIMA~1.2/share/maxima/5.31.2/share/draw/draw.lisp

(%i4) draw3d(enhanced3d = true, explicit(g(x,y),x,-4,4,y,-4,4));
```

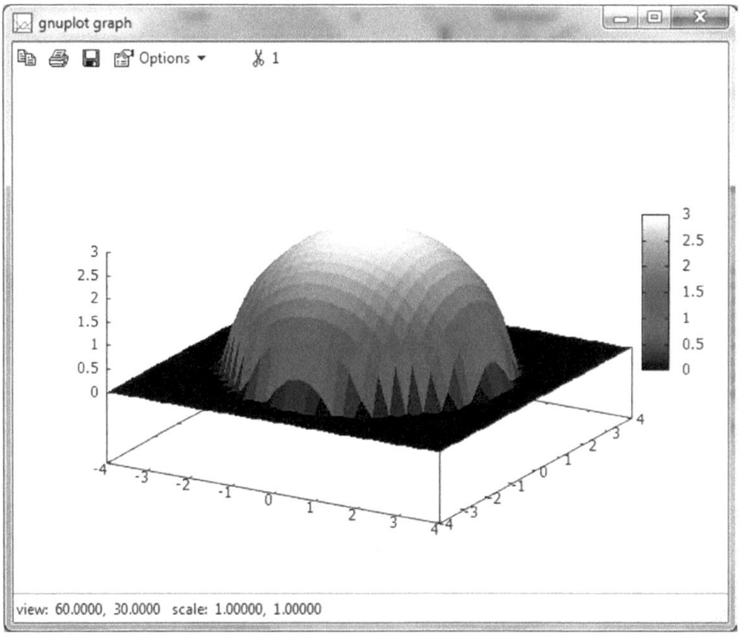

(See color insert.)

c) $f(x,y) = 6 - 3x - 2y$

```
(%i5) f(x,y):=6-3*x-2*y;
(%o5) f(x,y):=6-3 x+(-2) y

(%i6) load(draw);
(%o6) C:/PROGRA~1/MAXIMA~1.2/share/maxima/5.31.2/share/draw/draw.lisp

(%i7) draw3d(explicit(f(x,y),x,-4,4,y,-4,4));
```

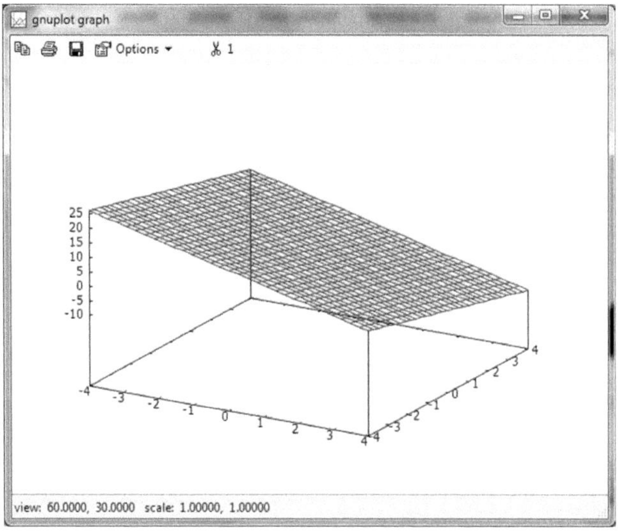


```
(%i8) f(x,y):=6-3*x-2*y;
(%o8) f(x,y):=6-3 x+(-2) y

(%i9) load(draw);
(%o9) C:/PROGRA~1/MAXIMA~1.2/share/maxima/5.31.2/share/draw/draw.lisp

(%i10) draw3d(enhanced3d = true,explicit(f(x,y),x,-4,4,y,-4,4));
```

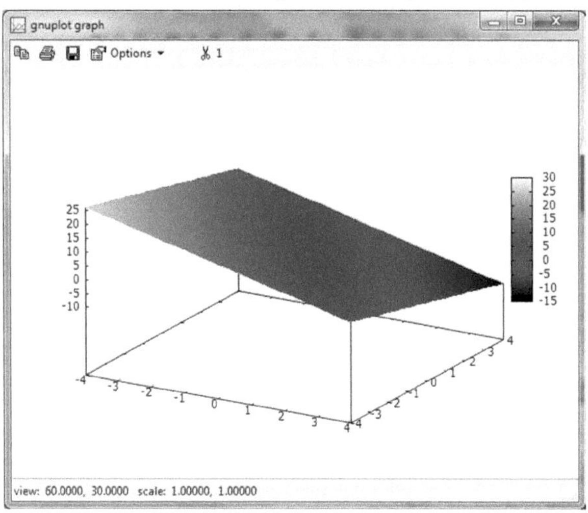

(See color insert.)

PARTIAL DERIVATIVES

We do partial derivatives in the natural way.

Find the indicated partial derivative(s)

1) $f(x, y) = x^4y^2 - x^3y; f_{xxx}$

```
(%i4) diff(x^4*y^2-x^3*y,x,3);
(%o4) 24 x y^2 -6 y
```

2) $f(x, y) = x^4y^2 - x^3y; f_{xyx}$

```
(%i6) diff(x^4*y^2-x^3*y,x,1,y,1,x,1);
(%o6) 24 x^2 y-6 x
```

3) $f(x, y, z) = e^{xyz^2}; f_{xyz}$

```
(%i10) diff(f,z,1,y,1,x,1);
(%o10) 0
```

CHAIN RULE

Suppose we have a function $f(x, y) = e^{x2}$ sin y

```
(%i1) f(x,y):= exp(x^2)*sin(y);
(%o1) f(x,y):=exp(x^2) sin(y)
```

We saw what the partial derivatives of f were with respect to x and y. But suppose x and y are functions of some other variables, for instance,

```
(%i2) [x,y]:[s^2*t,s*t^2];
```

$$[s^2\ t, s\ t^2]$$

What are the partial derivatives of *f* with respect to *s* and *t*? Maxima does the chain rule automatically.

```
(%i3) diff(f(x,y),s);
```

$$4\ s^3\ t^2\ \%e^{s^4\ t^2}\ \sin(s\ t^2) + t^2\ \%e^{s^4\ t^2}\ \cos(s\ t^2) \quad (\%o3)$$

```
(%i4) diff(f(x,y),t);
```

$$2\ s^4\ t\ \%e^{s^4\ t^2}\ \sin(s\ t^2) + 2\ s\ t\ \%e^{s^4\ t^2}\ \cos(s\ t^2) \quad (\%o4)$$

NB: The derivative with respect to *x* does not work anymore.
We could fix this by using different letters, for example *u* and v:

```
(%i5) diff(f(u,v),u);
```

$$2\ u\ \%e^{u^2}\ \sin(v) \quad (\%o5)$$

A better way to fix it is kill the relationship between *x* and *s*, *ts*,*t*.

```
(%i6) kill(x,y);
(%o6) done
```

```
(%i7) diff(f(x,y),x);
```

$$2\ x\ \%e^{x^2}\ \sin(y) \quad (\%o7)$$

```
(%i8) diff(f(x,y),y);
```

$$\%e^{x^2}\ \cos(y) \quad (\%o8)$$

Use the Chain Rule to find *dz*/*dtdz*/*dt*

1) $z = x^2 + y^2 + xy,\ x = sin\ t, y = e^t$

```
(%i1) f(x,y):= x^2+y^2+x*y;
(%o1) f(x,y):=x^2+y^2+x y

(%i2) [x,y]:[sin(t),e^t];
(%o2) [sin(t),e^t]

(%i3) diff(f(x,y),t);
(%o3) 2 cos(t) sin(t)+e^t log(e) sin(t)+e^t cos(t)+2 e^(2 t) log(e)
```

2) $z = cos(x + 4y)\ x = 5t4,\ y = 1/4$

```
(%i1) f(x,y):= x^2+y^2+x*y;
(%o1) f(x,y):=x^2+y^2+x y

(%i2) [x,y]:[sin(t),e^t];
(%o2) [sin(t),e^t]

(%i3) diff(f(x,y),t);
(%o3) 2 cos(t) sin(t)+e^t log(e) sin(t)+e^t cos(t)+2 e^(2 t) log(e)
```

DIRECTIONAL DERIVATIVES AND THE GRADIENT

Find the Gradient and Directional Derivative of the following functions

a) $f(x,y) = e^{x2}$. Directional derivative at the point (1,2) in the direction of the vector v =<3,4>

```
(%i1) f(x,y):=exp(x^2)*sin(y);
(%o1) f(x,y):=exp(x^2) sin(y)

(%i2) load(vect);
(%o2) C:/PROGRA~1/MAXIMA~1.2/share/maxima/5.31.2/share/vector/vect.mac

(%i3) scalefactors([x,y]);
(%o3) done

(%i7) gdf: grad(f(x,y));
(%o7) grad(%e^(x^2) sin(y))

(%i8) ev(express(gdf),diff);
(%o8) [2 x %e^(x^2) sin(y), %e^(x^2) cos(y)]

(%i10) define(gdf(x,y),%);
(%o10) gdf(x,y):=[2 x %e^(x^2) sin(y), %e^(x^2) cos(y)]
```

The directional derivative of *f* at the point (1,2) in the direction vector v = <3,4> is

```
(%i11) v:[3,4];
(%o11) [3,4]

(%i12) (gdf(1,2).v)/sqrt(v.v);
(%o12) (6 %e sin(2)+4 %e cos(2))/5

(%i13) ev(%,diff);
(%o13) (6 %e sin(2)+4 %e cos(2))/5

(%i14) float(%);
(%o14) 2.061108499400332
```

b) $f(x,y) = x^2y^3 - 4y$ at the point (2,–1) in the direction of the vector v = <2,5>

```
(%i1) f(x,y):=x^2*y^3-4*y;
(%o1) f(x,y):=x^2 y^3-4 y

(%i2) load(vect);
(%o2) C:/PROGRA~1/MAXIMA~1.2/share/maxima/5.31.2/share/vector/vect.mac

(%i3) scalefactors([x,y]);
(%o3) done

(%i4) gdf: grad(f(x,y));
(%o4) grad(x^2 y^3-4 y)

(%i5) ev(express(gdf),diff);
(%o5) [2 x y^3,3 x^2 y^2-4]

(%i6) define(gdf(x,y),%);
(%o6) gdf(x,y):=[2 x y^3,3 x^2 y^2-4]

(%i7) v:[2,5];
(%o7) [2,5]

(%i8) (gdf(2,-1).v)/sqrt(v.v);
(%o8) 32/sqrt(29)

(%i9) ev(%,diff);
(%o9) 32/sqrt(29)

(%i10) float(%);
(%o10) 5.94225082166566
```

LOCAL EXTREMA

To find critical points we need to solve the system of equations $f_x = 0$ and $f_y = 0$.

1) $f(x,y) = 2x^4 + 2y^4 - 8xy$

```
(%i1) f(x,y):=2*x^4+2*y^4-8*x*y;
(%o1) f(x,y):=2 x^4+2 y^4+(-8) x y

(%i2) fx:diff(f(x,y),x);
(%o2) 8 x^3-8 y

(%i3) fy:diff(f(x,y),y);
(%o3) 8 y^3-8 x

(%i4) solve([fx,fy],[x,y]);
(%o4) [[x=(-1)^(1/4) %i,y=-(-1)^(3/4) %i],[x=-(-1)^(1/4),y=-(-1)^(3/4)],[x=-(-1)^(1/4) %i,y=(-1)^(3/4)
%i],[x=(-1)^(1/4),y=(-1)^(3/4)],[x=-%i,y=%i],[x=%i,y=-%i],[x=-1,y=-1],[x=1,y=1],[
x=0,y=0]]
```

Critical points are at (–1,–1), (1,1), (0,0).

```
(%i7) f(-1,-1);
(%o7) -4

(%i8) f(0,0);
(%o8) 0

(%i9) f(1,1);
(%o9) -4
```

We conclude that (1,1) and (–1,–1) are a local minimum of f, and the value is –4.

(0,0) is a saddle point.

2) $f(x) = x^4 + y^4 - 4xy + 1$

```
(%i10) f(x,y):=x^4+y^4-4*x*y+1;
(%o10) f(x,y):=x^4+y^4+(-4) x y+1

(%i11) fx:diff(f(x,y),x);
(%o11) 4 x^3-4 y

(%i12) fy:diff(f(x,y),y);
(%o12) 4 y^3-4 x

(%i13) solve([fx,fy],[x,y]);
(%o13) [[x=(-1)^(1/4) %i,y=-(-1)^(3/4) %i],[x=-(-1)^(1/4),y=-(-1)^(3/4)],[x=-(-1)^(1/4) %i,y=(-1)^(3/4) %i],[x=(-1)^(1/4),y=(-1)^(3/4)],[x=-%i,y=%i],
[x=%i,y=-%i],[x=-1,y=-1],[x=1,y=1],[x=0,y=0]]
```

```
(%i14) f(-1,-1);
(%o14) -1
```

```
(%i15) f(0,0);
(%o15) 1
```

```
(%i16) f(1,1);
(%o16) -1
```

We conclude that (1,1) and (–1,–1) are a local minimum of f, and the value is –1.

(0,0) is a saddle point.

LAGRANGE MULTIPLIERS

1) Find the extreme values of the function $f(x,y) = 2x^2 + y^2$ on the circle $x^2 + y^2 = 1$

```
(%i1) f(x,y):=2*x^2+y^2;
```

(%o1) $f(x,y):=2\,x^2+y^2$

```
(%i2) g:x^2+y^2;
```

(%o2) y^2+x^2

We set up the system of equations *grad*(*f*) = *h***grad*(*g*), *g* = 1. We don't use "lambda" because the name is already reserved for an existing function in Maxima.

```
(%i3) grad(f) = h*grad(g),g = 1;
(%o3) grad(f)=grad(1)h

(%i4) eq1: diff(f(x,y),x)=h*diff(g,x);
(%o4) 4 x=2 h x

(%i5) eq2:diff(f(x,y),y)=h*diff(g,y);
(%o5) 2 y=2 h y

(%i6) eq3: g=1;
(%o6) y^2+x^2=1
```

Now we solve the system for x, y, and h.

```
(%i7) solve([eq1,eq2,eq3],[x,y,h]);
(%o7) [[x=1,y=0,h=2],[x=-1,y=0,h=2],[x=0,y=-1,h=1],[x=0,y=1,h=1]]
```

We see that the extreme values lie among (1,0),(–1,0),(0,–1),(0,1).

```
(%i8) [f(1,0),f(-1,0),f(0,-1),f(0,1)];
(%o8) [2,2,1,1]
```

So the minima occur at (1,0) and (–1,0); the maxima occur at (0,–1) and (0,1).

2) Find the extreme values of the function $f(x,y) = 2x^2 + y^2$ on the circle $x^2 + y^2 = 1$

```
(%i9) f(x,y):=x^2+2*y^2;
(%o9) f(x,y):=x^2+2 y^2

(%i10) g:x^2+y^2;
(%o10) y^2+x^2

(%i11) grad(f) = h*grad(g),g = 1;
(%o11) grad(f)=grad(1) h

(%i12) eq1: diff(f(x,y),x)=h*diff(g,x);
(%o12) 2 x=2 h x

(%i13) eq2:diff(f(x,y),y)=h*diff(g,y);
(%o13) 4 y=2 h y

(%i14) eq3: g=1;
(%o14) y^2+x^2=1

(%i15) solve([eq1,eq2,eq3],[x,y,h]);
(%o15) [[x=1,y=0,h=1],[x=-1,y=0,h=1],[x=0,y=-1,h=2],[x=0,y=1,h=2]]

(%i16) [f(1,0),f(-1,0),f(0,-1),f(0,1)];
(%o16) [1,1,2,2]
```

So the minima occur at (1,0) and (–1,0); the maxima occur at (0,–1) and (0,1).

APPLICATION

1) Sketch the graph of:
   a) $f(x,y) = 4x^2 + y^2$
   b) $g(x,y) = y^2 + 1$
   c) $f(x,y) = 4x^2 + y^2 + 1$

2) Find the indicated partial derivative(s)
   a) $g(r,s,t) = e^r \sin(st)$; g_{rst}
   b) $f(x,y) = 4x^2 + y^2$; $fxyx$
   c) $f(x,y,z) = e^{xyz2}$; f_{xxyz}

3) Use the Chain Rule to find dz/dt
   a) $z = \sqrt{1+x^2+y^2}, x = \ln t, y = \cos t$
   b) $z = xe^{y/z}$ $x = e^t, y = 1 - t, z = 1 + 2t$
   c) $z = x^2y^3$ $x = s \cos t, y = 1 - 2\,st$
4) Find the directional derivative of the function at the given point in the direction of the vector v
   a) $f(x,y) = e^x \sin y$, $(0, \pi/3)$, $v = <-6,8>$
   b) $f(x,y) = \frac{x}{x^2+y^2}$, $(1,2)$, $v = <3,5>$
   c) $g(p,q) = p^4 - p^2q^3$, $(2,1)$, $v = <1,3>$
5) Find and classify the critical points of the following functions
   a) $f(x,y) = 10x^2y - 5x^2 - 4y^2 - x^4 - 2y^4$
   b) $f(x,y) = 4 + x^3 + y^3 - 3xy$
   c) $f(x,y) = y\cos x$
6) Find the extreme values of the function $f(x,y) = x^2 + y^2$ on the circle $x^2 + y^2 = 1$
7) Find the extreme values of the function $f(x,y) = 3y + x$ on the circle $x^2 + y^2 = 10$
8) Find the extreme values of the function $f(x,y) = -x^2 + y^2$ on the circle
   $\frac{1}{4}x^2 + y^2 = 1$

3.10 MULTIPLE INTEGRALS

DOUBLE INTEGRALS

Find the integrals using Maxima

1) $\iint x^3 - 3xy\ dydx$

```
(%i1) f(x,y):=x^3-3*x*y;
```

(%o1) $f(x,y):=x^3-3\,x\,y$

```
(%i2) integrate(integrate(f(x,y),y),x);
```

(%o2) $\frac{x^4\,y}{4}-\frac{3\,x^2\,y^2}{4}$

2) $\iint x^2 - y\, dydx$

```
(%i3) f(x,y):=x^2-y;
```

(%o3) $f(x,y):=x^2-y$

```
(%i4) integrate(integrate(f(x,y),y),x);
```

(%o4) $\frac{x^3\,y}{3}-\frac{x\,y^2}{2}$

3) $\iint x-3y^2\, dydx$

```
(%i5) f(x,y):=x-3*y^2;
```

(%o5) $f(x,y):=x-3\,y^2$

```
(%i6) integrate(integrate(f(x,y),y),x);
```

(%o6) $\frac{x^2\,y}{2}-x\,y^3$

3) $\int_0^1 \int_{\sqrt{x}}^{2-x} \left(x^3 - 3xy\right) dydx$

```
(%i7) f(x,y):=x^3-3*x*y;
(%o7) f(x,y):=x^3 -3 x y

(%i8) integrate(integrate(f(x,y),y,x^1/2,2-x),x,0,1);
        173
(%o8) - ---
        160
```

5) $\int_0^1 \int_{\sqrt{x}}^{2-x} \left(x-3y^2\right) dydx$

```
(%i9) f(x,y):=x-3*y^2;
(%o9) f(x,y):=x -3 y^2

(%i10) integrate(integrate(f(x,y),y,x^1/2,2-x),x,0,1);
         103
(%o10) - ---
          32
```

6) $\int_0^2 \int_1^2 \left(x-3y^2\right) dydx$

```
(%i9) f(x,y):=x-3*y^2;
(%o9) f(x,y):=x -3 y^2

(%i11) integrate(integrate(f(x,y),y,1,2),x,0,2);
(%o11) -12
```

DOUBLE INTEGRATION IN POLAR COORDINATES

We simply make the substitution $x = rcos(theta)$ and $y = rsin(theta)$

$$\int_{-\pi/2}^{\pi/2} \int_{0}^{2\cos\theta} (x^2 + y^2) dA$$

```
(%i1) f(x,y):=x^2+y^2;
(%o1) f(x,y):=x^2+y^2

(%i2) [x,y]:[r*cos(theta),r*sin(theta)];
(%o2) [r cos(θ), r sin(θ)]

(%i3) integrate(integrate(f(x,y)*r,r,0,2*cos(theta)),theta,-%pi/2,%pi/2);
(%o3) 3π/2
```

$$\int_0^{\pi}\int_1^{2} 3x+4y^2 dA$$

```
(%i4) f(x,y):=3*x+4*y^2;
(%o4) f(x,y):=3 x+4 y^2

(%i5) [x,y]:[r*cos(theta),r*sin(theta)];
(%o5) [r cos(θ), r sin(θ)]

(%i6) integrate(integrate(f(x,y)*r,r,1,2),theta,0,%pi);
(%o6) 15π/2
```

$$\int_0^{2\pi}\int_0^{1} 1-x^2-y^2 dA$$

```
(%i7) f(x,y):=1-x^2-y^2;
(%o7) f(x,y):=1-x^2-y^2

(%i8) [x,y]:[r*cos(theta),r*sin(theta)];
(%o8) [r cos(θ), r sin(θ)]

(%i9) integrate(integrate(f(x,y)*r,r,0,1),theta,0,2*%pi);
(%o9) π/2
```

TRIPLE INTEGRALS

Triple integrals are done just like double integrals.

$$\int_0^1\int_0^{-x}\int_0^{x+y} x^2yz\ dzdydx$$

```
(%i4) integrate(integrate(integrate (x^2*y*z,z,0,x+y),y,0,-x),x,0,1);
(%o4) 1/168
```

$$\int_0^1\int_0^{1-x}\int_0^{1-x-y} z\ dzdydx$$

```
(%i5) integrate(integrate(integrate (z,z,0,1-x-y),y,0,1-x),x,0,1);
(%o5) 1/24
```

$$\int_0^1\int_x^{2x}\int_0^{y} 2xyz\ dzdydx$$

```
(%i6) integrate(integrate(integrate (2*x*y*z,z,0,y),y,x,2*x),x,0,1);
(%o6) 5/8
```

TRIPLE INTEGRALS IN CYLINDRICAL COORDINATES

This integral is computed just like the triple integrals except we multiply the integrand by r.

$$\int_{-2}^2\int_0^{\sqrt{4-x^2}}\int_0^{3} yz\ dzdydx$$

Note that r goes from 0 to 2 and θ goes from 0 to π

```
(%i1) f(x,y,z):=y*z;
(%o1) f(x,y,z):=y z

(%i2) [x,y,z]:[r*cos(theta),r*sin(theta),z];
(%o2) [r cos(θ), r sin(θ), z]

(%i3) integrate(integrate(integrate(f(x,y,z)*r,z,0,3),r,0,2),theta,0,%pi);
(%o3) 24
```

$$\int_{-2}^{2}\int_{-\sqrt{4-x^2}}^{\sqrt{4-x^2}}\int_{\sqrt{x^2+y^2}}^{2}(x^2+y^2)dzdydx$$

Note that r goes from 0 to 2 and θ goes from 0 to 2π and z from r to 2

```
(%i4) f(x,y,z):=x^2+y^2;
(%o4) f(x,y,z):=x^2+y^2

(%i2) [x,y,z]:[r*cos(theta),r*sin(theta),z];
(%o2) [r cos(θ), r sin(θ), z]

(%i6) integrate(integrate(integrate(f(x,y,z)*r,z,r,2),r,0,2),theta,0,2*%pi);
       16 π
(%o6)  ----
        5
```

TRIPLE INTEGRALS IN SPHERICAL COORDINATES

This integral is computed just like the triple integrals except we multiply the integrand by $\rho^2 \sin(\varphi)$

$$\int_{-1}^{1}\int_{0}^{\sqrt{1-x^2}}\int_{0}^{\sqrt{1-x^2-y^2}} xzdzdydx$$

Note that ρ goes from 0 to 1, θ goes from 0 to π and φ goes from 0 to $\pi/2\pi//2$.

```
(%i1) f(x,y,z):=x*z;
(%o1) f(x,y,z):=x z

(%i2) [x,y,z]:[rho*sin(phi)*cos(theta),rho*sin(phi)*sin(theta),rho*cos(theta)];
(%o2) [sin(ϕ)ρ cos(θ), sin(ϕ)ρ sin(θ), ρ cos(θ)]

(%i3) integrate(integrate(integrate(f(x,y,z)*rho^2*sin(phi),rho,0,1),theta,0,%pi),phi,0,%pi/2);
(%o3) π^2/40
```

APPLICATION

1. Calculate the iterated integral

a) $\int_{1}^{4}\int_{0}^{2}\left(6x^2y-2x\right)dydx$

b) $\int_{0}^{1}\int_{1}^{2}\left(4x^3-9x^2y^2\right)dydx$

c) $\int_{1}^{4}\int_{1}^{2}\left(\frac{x}{y}+\frac{y}{x}\right)dydx$

d) $\int_{-\pi/2}^{\pi/2}\int_{0}^{2cos\ \theta} x^2+y^2dA$

e) $\int_{-3}^{3}\int_{0}^{\sqrt{9-x^2}} \sin(x^2+y^2dA$

f) $\int_{0}^{2}\int_{0}^{z^2}\int_{0}^{y-z}\left(2x-y\right)dxdydz$

g) $\int_{0}^{1}\int_{0}^{-x}\int_{0}^{y}x^2yz\ dzdydx$

h) $\int_{1}^{2}\int_{0}^{2z}\int_{0}^{\ln x}xe^{-y}dydxdz$

2. Evaluate the integral by changing to cylindrical coordinates

a) $\int_{-2}^{2}\int_{-\sqrt{4-y^2}}^{\sqrt{4-y^2}}\int_{\sqrt{x^2+y^2}}^{2} xz\ dzdydx$

b) $\int_{-3}^{3}\int_{0}^{\sqrt{9-x^2}}\int_{0}^{9-x^2-y^2}\sqrt{x^2+y^2}\,dzdydx$

3. Evaluate the integral by changing to spherical coordinates

a) $\int_{0}^{1}\int_{0}^{\sqrt{1-x^2}}\int_{\sqrt{x^2+y^2}}^{\sqrt{2-x^2-y^2}} xy\ dzdydx$

b) $\int_{-a}^{a}\int_{-\sqrt{a^2-y^2}}^{\sqrt{a^2-y^2}}\int_{-\sqrt{a^2-x^2-y^2}}^{\sqrt{a^2-x^2-y^2}}(x^2z+y^2z+z^3)dzdxdy$

c) $\int_{-2}^{2}\int_{-\sqrt{4-x^2}}^{\sqrt{4-x^2}}\int_{2-\sqrt{4-x^2-y^2}}^{2+\sqrt{4-x^2-y^2}}\left(x^2+y^2+z^2\right)^{3/2}dzdydx$

3.11 VECTOR CALCULUS

VECTOR FIELDS

Plot the following vector fields using maxima

$F(x,y) = < \cos y, x >$

```
load(draw)
coord:setify(makelist(k,k,-4,4));
points2d: listify(cartesian_product(coord,coord));
vf2d(x,y):= vector([x,y],[cos(y),x]/6);
vect2:makelist(vf2d(k[1],k[2]),k,points2d);
apply(draw2d, append([color = blue],vect2));
```

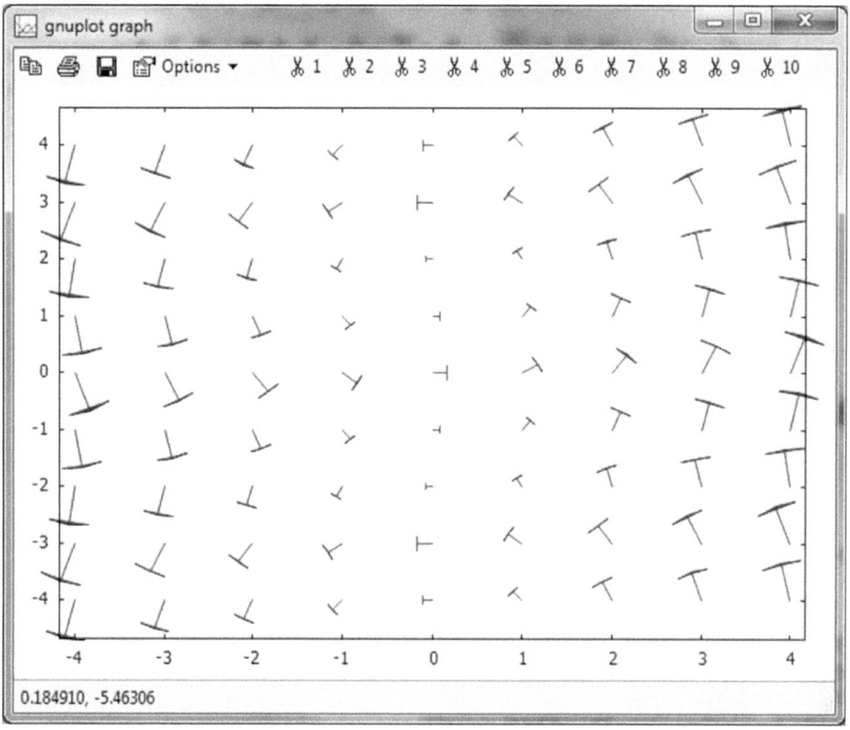

$F(x,y) = <-y, x>$

```
(%i1) load(draw);
Loading C:/Users/pauly.awad/maxima/binary/binary-gcl/share/draw/grcommon.o
Finished loading C:/Users/pauly.awad/maxima/binary/binary-gcl/share/draw/grcommon.o
Loading C:/Users/pauly.awad/maxima/binary/binary-gcl/share/draw/gnuplot.o
Finished loading C:/Users/pauly.awad/maxima/binary/binary-gcl/share/draw/gnuplot.o
Loading C:/Users/pauly.awad/maxima/binary/binary-gcl/share/draw/vtk.o
Finished loading C:/Users/pauly.awad/maxima/binary/binary-gcl/share/draw/vtk.o
Loading C:/Users/pauly.awad/maxima/binary/binary-gcl/share/draw/picture.o
Finished loading C:/Users/pauly.awad/maxima/binary/binary-gcl/share/draw/picture.o
(%o1) C:/PROGRA~1/MAXIMA~1.2/share/maxima/5.31.2/share/draw/draw.lisp

(%i2) coord:setify(makelist(k,k,-4,4));
      points2d: listify(cartesian_product(coord,coord));
      vf2d(x,y):= vector([x,y],[-y,x]/6);
      vect2:makelist(vf2d(k[1],k[2]),k,points2d);
      apply(draw2d, append([color = blue],vect2));
```

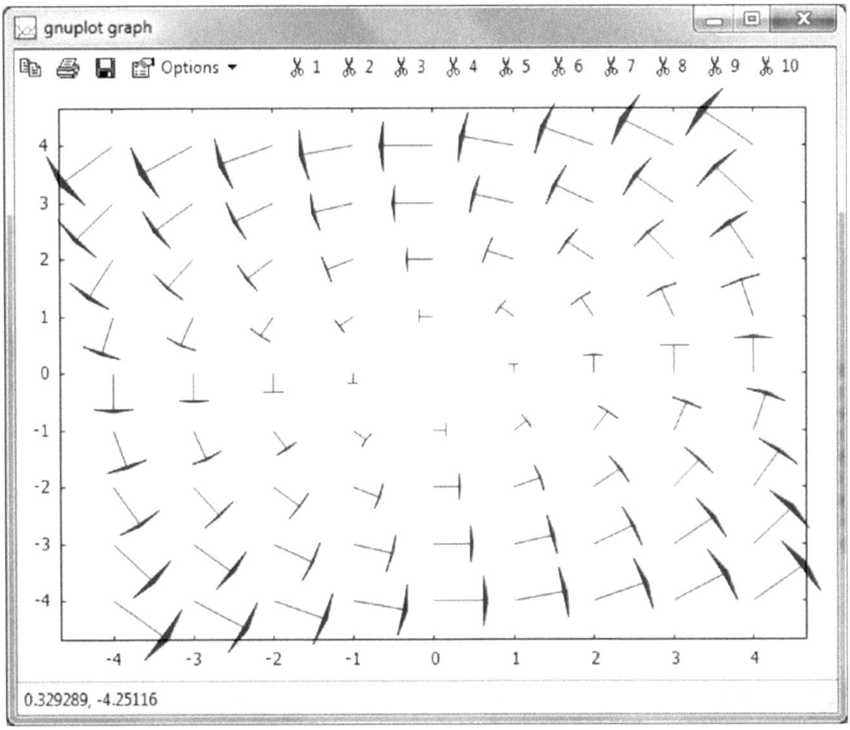

$F(x,y) = < y, \sin x>$

```
load(draw);
coord:setify(makelist(k,k,-4,4));
points2d: listify(cartesian_product(coord,coord));
vf2d(x,y):= vector([x,y],[-y,x]/6);
vect2:makelist(vf2d(k[1],k[2]),k,points2d);
apply(draw2d, append([color = blue],vect2));
```

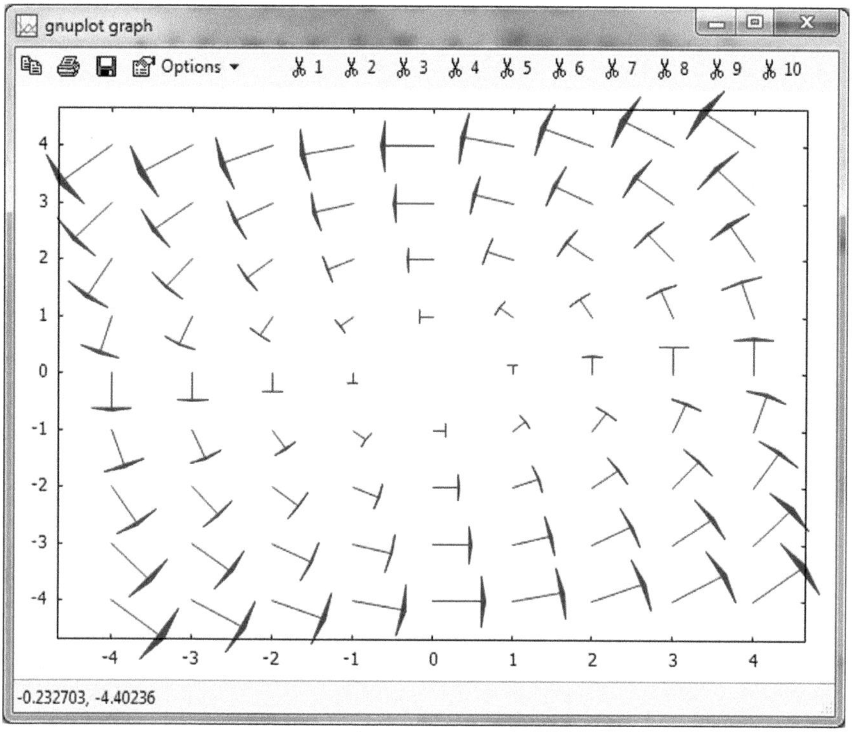

$F(x, y, z) = < z, x, y>$

```
load(draw);
coord:setify(makelist(k,k,-4,4));
points3d: listify(cartesian_product(coord,coord,coord));
vf3d(x,y,z):= vector([x,y,z],[z,x,y]/6);
vect3:makelist(vf3d(k[1],k[2],k[3]),k,points3d);
apply(draw3d, append([color = blue],vect3));
```

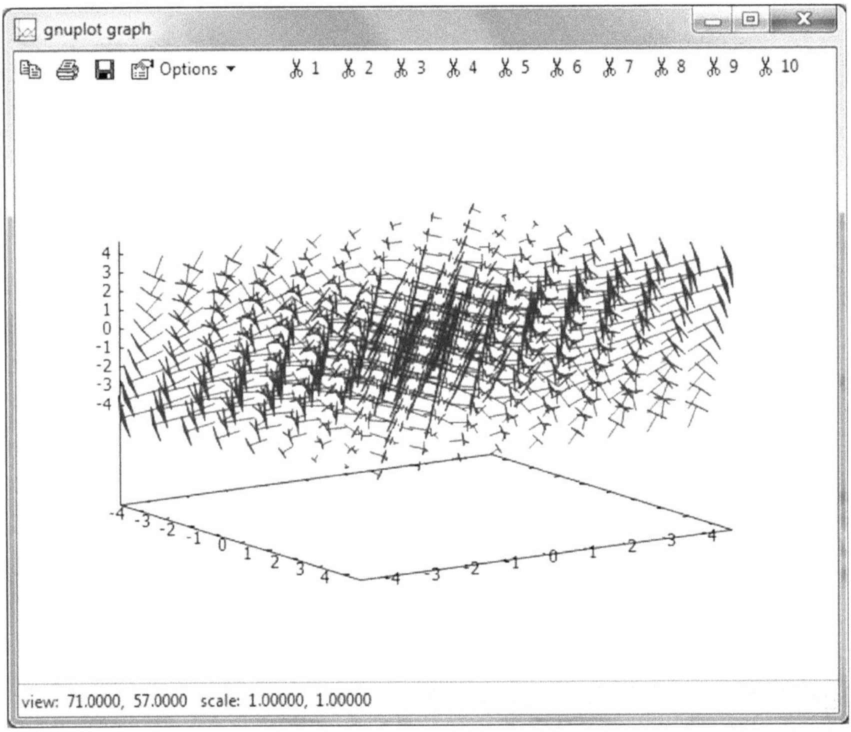

$F(x, y, z) = \langle y, -2, x \rangle$

```
load(draw);
coord:setify(makelist(k,k,-4,4));
points3d: listify(cartesian_product(coord,coord,coord));
vf3d(x,y,z):= vector([x,y,z],[y,-2,x]/8);
vect3:makelist(vf3d(k[1],k[2],k[3]),k,points3d);
apply(draw3d, append([color = blue],vect3));
```

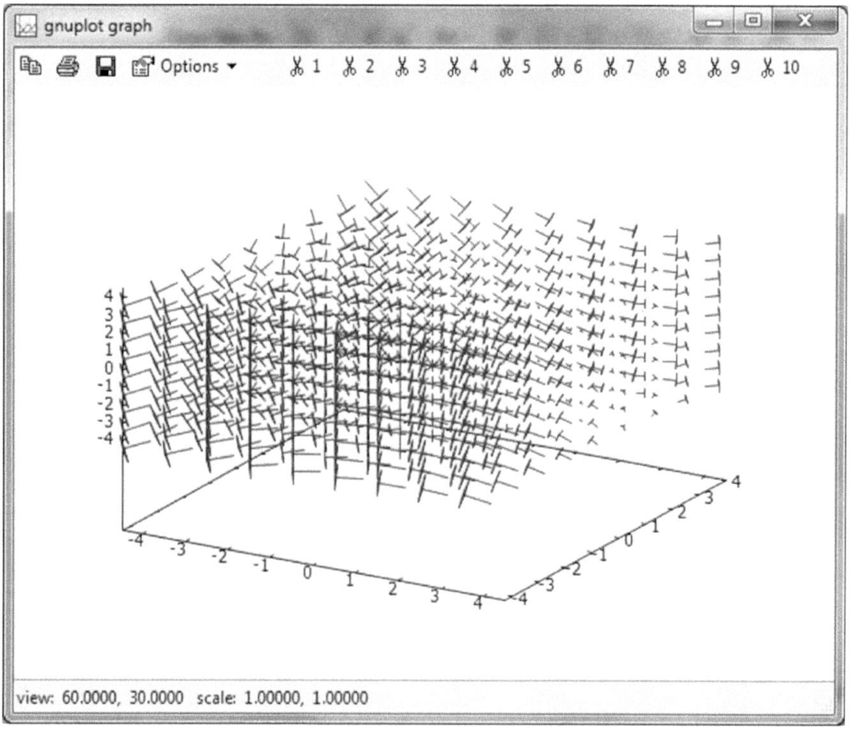

GRADIENT VECTOR FIELDS

We can plot this vector filed in two ways:

1) With the **draw** package
2) With the **ploteq** function

$f(x,y) = x^2 - y^2$

```
(%i13) ploteq(-(x^2+y^2),[x,y],[x,-4,4],[y,-4,4],[vectors, "blue"]);
```

NB: We should remember to put a minis sign in front of the original function (ploteq plots vectors that are the opposite of gradient vectors).

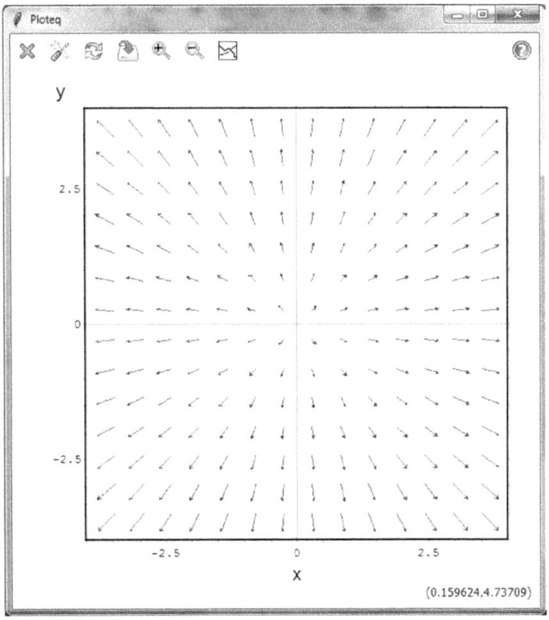

$f(x,y) = x^2y - y^3$

```
(%i1) ploteq(-(x^2*y-y^3),[x,y],[x,-4,4],[y,-4,4],[vectors, "blue"]);
```

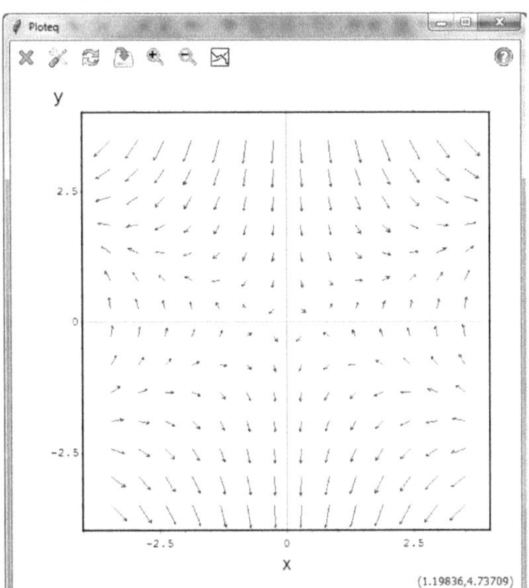

Line integrals with respect to arc length

1) $f(x,y) = x^2 + y^2$

We will integrate along the curve C parameterized by *cos t, sin* 2*t*> for $0 \leq t \leq 10 \leq t \leq 1$

```
(%i1) f(x,y):= x^2 +y^2;
(%o1) f(x,y):=x^2+y^2

(%i3) [x,y]:[cos(t), sin (2*t)];
(%o3) [cos(t),sin(2 t)]

(%i4) rp: diff([x,y], t);
(%o4) [-sin(t),2 cos(2 t)]

(%i5) romberg(f(x,y)*sqrt(rp.rp),t,0,1);
(%o5) 1.635879048260743
```

2) $f(x,y) = y^2 + x$

We will integrate along the curve C parameterized by <cos t, sin 2t > for $0 \leq t \leq 1$

```
(%i6) f(x,y):= x^2 +y;
(%o6) f(x,y):=x^2+y

(%i7) [x,y]:[cos(t), sin (2*t)];
(%o7) [cos(t),sin(2 t)]

(%i8) rp: diff([x,y], t);
(%o8) [-sin(t),2 cos(2 t)]

(%i9) romberg(f(x,y)*sqrt(rp.rp),t,0,1);
(%o9) 1.814053767365267
```

3) $f(x,y) = y^3$ $x = t^3, y = t, 0 \leq t \leq 2$

```
(%i10) f(x,y):= y^3;
(%o10) f(x,y):=y^3

(%i11) [x,y]:[t^3, t];
(%o11) [t^3,t]

(%i12) rp: diff([x,y], t);
(%o12) [3 t^2,1]

(%i13) romberg(f(x,y)*sqrt(rp.rp),t,0,2);
(%o13) 32.31497565793021
```

Line integrals of vector fields

1) $F(x,y,z) = <-xy^3, xz, yz^2>$

We will integrate along the curve parameterized $<t^2, t^3, t^4>$ for $0 \leq t \leq 1$

```
(%i1) F(x,y,z):= [-x*y^3,x*z,y*z^2];
(%o1) F(x,y,z):=[(-x) y^3,x z,y z^2]

(%i2) [x,y,z]:[t^2,t^3,t^4];
(%o2) [t^2,t^3,t^4]

(%i3) romberg(F(x,y,z).diff([x,y,z],t),t,0,1);
(%o3) 0.44615384616036
```

2) $F(x, y, z) = < < xy, yz, zx >$
We will integrate along the quarter-circle $< t, t^2, t^3 >$ for $0 \leq t \leq 1$

```
(%i13) F(x,y,z):= [x*y,y*z,z*x];
(%o13) F(x,y,z):=[x y,y z,z x]

(%i14) [x,y,z]:[t,t^2,t^3];
(%o14) [t,t^2,t^3]

(%i15) romberg(F(x,y,z).diff([x,y,z],t),t,0,1);
(%o15) 0.96428571428571
```

APPLICATION

1) Plot the following vector fields using maxima
   a) $F(x, y) = < -y, 2x >$
   b) $F(x, y) = < 0.3, -0.4 >$
   c) $F(x, y) = < -1/2, (y - x) >$
   d) $F(x, y, z) = < -1, x, 0>$
   e) $F(x, y, z) = < 0,0, - y >$
   f) $F(x, y, z) = < 2x, z, -y >$
2) Plot the following gradient vector fields using maxima
   a) $f(x, y) = \ln(1 + x^2 + 2y^2)$
   b) $f(x,y) = xe^{xy}$
   c) $f(x,y,z) = \sqrt{x^2 + y^2 + z^2}$
   d) $f(x, y) = x^2 - y$
3) Evaluate the line integral, where C is the given curve.
   a) $\int_c xyds$, $C: x = t^2, y = 2t, 0 \le t \le 1$
   b) $\int_c x\sqrt{y}ds$, $C: x = t^2, y = 2t, 0 \le t \le 1$
   c) $\int_c xyz\, ds$, $C: x = t^2, y = 2t, z = t^4, 0 \le t \le 1$
4) Evaluate the integral where C is given by the vector function $r(t)$
   a) $F(x, y) = <xy, 3y^2>$, $r(t) = < 11t^4, t^3 >$, $0 < t < 1$
   b) $F(x, y, z) = < x + y, y - z, z^2 <$, $r(t) = < t^2, t^3, t^2 >$, $0 < t < 1$
   c) $F(x, y, z) = < \sin x, \cos y, xz >$, $r(t) = < t^3, - t^2, t >$, $0 \le t \le 1$

CHAPTER 4

LINEAR ALGEBRA

4.1 INTRODUCTION TO MATRICES AND LINEAR ALGEBRA

Maxima has many functions for defining and manipulating matrices and these functions can be used without loading in any additional packages. Here is the list (but several of these – see below – need separate packages loaded) from the Maxima HTML Help Manual (In the Help Manual index, type: matrix, and then click on the category: Matrices, to see this list.)

```
Category: Matrices

addcol  addrow  adjoint  augcoefmatrix  cauchy_matrix  charpoly  coefmatrix
col  columnvector  covect  copymatrix  determinant  detout  diag  diagmatrix
doallmxops  domxexpt  domxmxops  domxnctimes  doscmxops  doscmxplus
echelon  eigen  ematrix  entermatrix  genmatrix  ident  invert  list_matrix_entries
lmxchar  matrix  matrix_element_add  matrix_element_mult
matrix_element_transpose  matrixmap  matrixp  mattrace  minor  ncharpoly
newdet  nonscalar  nonscalarp  permanent  rank  ratmx  row  scalarmatrixp
scalarp  setelmx  sparse  submatrix  tracematrix  transpose  triangularize  zeromatrix
```

4.2 FUNCTIONS IN THE *ALGEBRA* MENU

The items in the *Algebra* menu, shown in the figure to the right, are presented in the following sections.

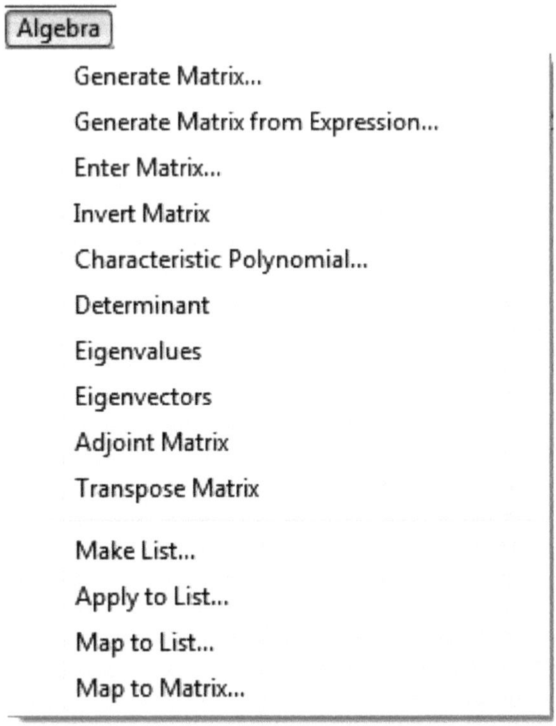

GENERATE MATRIX

The *Algebra > Generate matrix.* utilizes a function of the matrix sub-indices *i* and *j*, defined previously to invoking the menu item

```
(%i1) f[i,j]:=1/(i+j);
```

(%o1) $f_{i,j} := \frac{1}{i+j}$

Then, use *f* in the dialogue form
Go to Algebra → Generate matrix

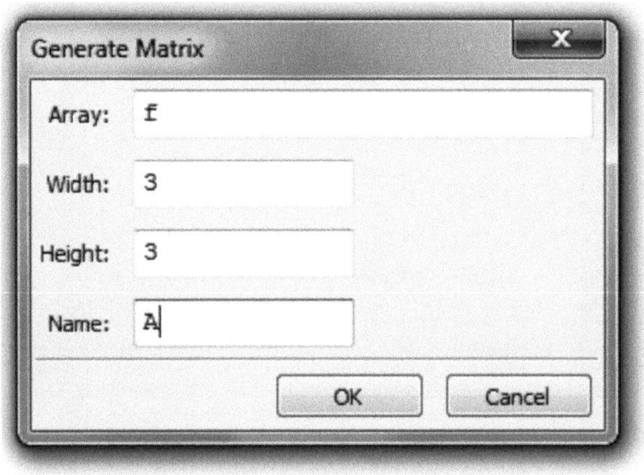


```
(%i3) A: genmatrix(f, 3, 3);
```

(%o3) $$\begin{bmatrix} \frac{1}{2} & \frac{1}{3} & \frac{1}{4} \\ \frac{1}{3} & \frac{1}{4} & \frac{1}{5} \\ \frac{1}{4} & \frac{1}{5} & \frac{1}{6} \end{bmatrix}$$

ENTER MATRIX

The *Enter matrix menu* item is used to enter a matrix of given dimensions. The resulting dialogue form provides the following options:

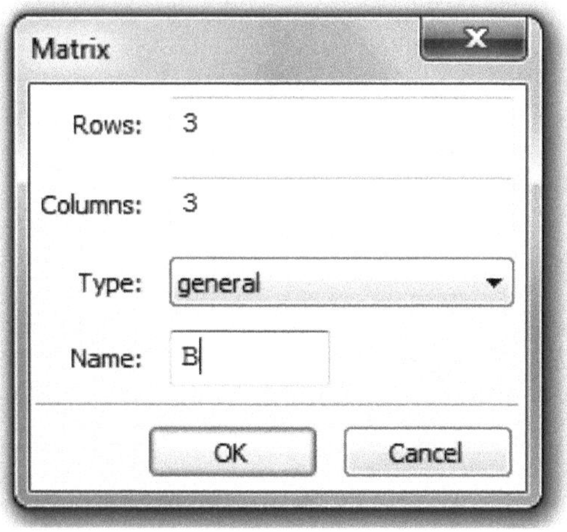

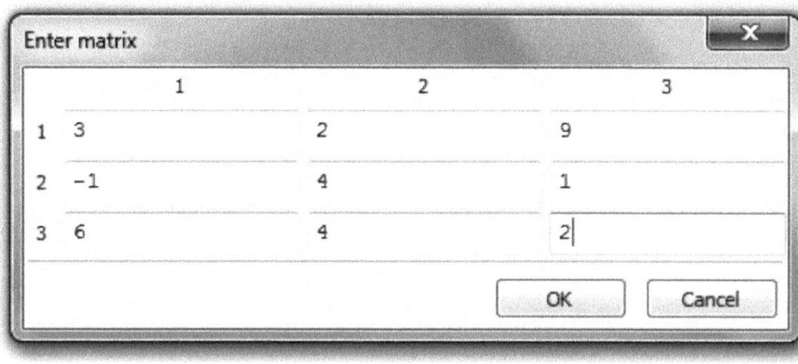


```
(%i4) B: matrix(
        [3,2,9],
        [-1,4,1],
        [6,4,2]
      );
```

(%o4) $\begin{bmatrix} 3 & 2 & 9 \\ -1 & 4 & 1 \\ 6 & 4 & 2 \end{bmatrix}$

So we just need to write the dimension of the matrix and define each element of the matrix

Enter the following matrices using Maxima:

a) C = [2]

```
(%i5) C: matrix(
        [2]
       );
```

(%o5) $\begin{bmatrix} 2 \end{bmatrix}$

b) [1, 0, 3, ½]

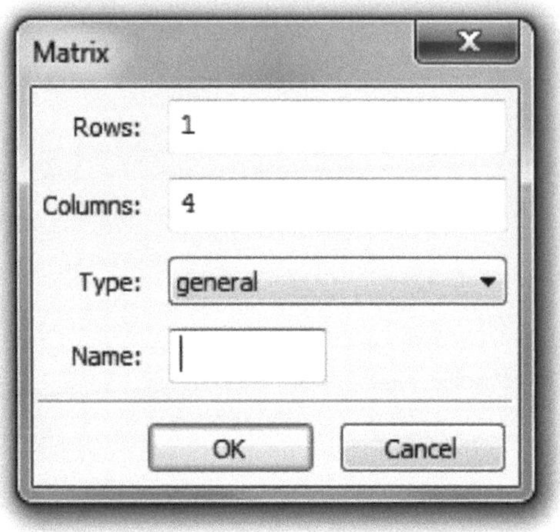

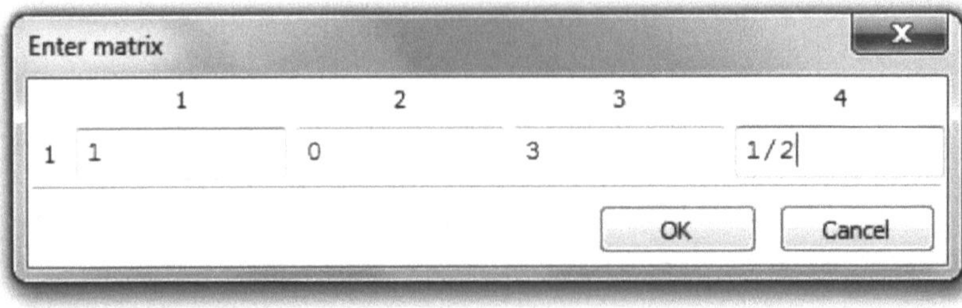


```
(%i6) matrix(
       [1,0,3,1/2]
      );
```

$$(\%o6)\quad \begin{bmatrix} 1 & 0 & 3 & \frac{1}{2} \end{bmatrix}$$

c) $\begin{bmatrix} 5 & 0 \\ 2 & -2 \end{bmatrix}$

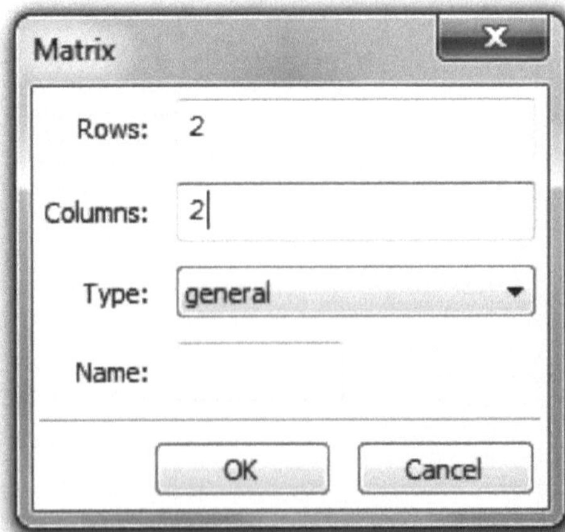

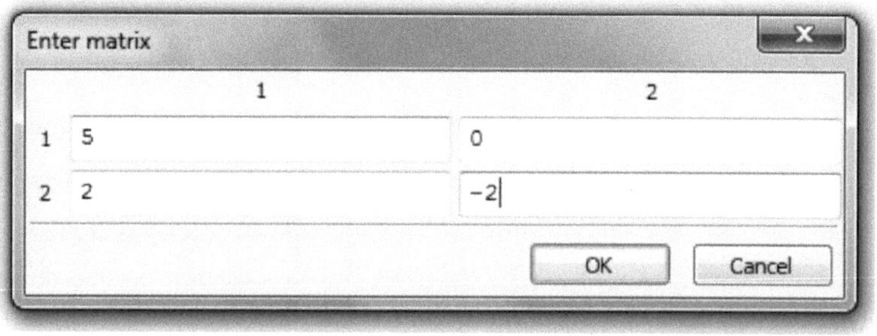


```
(%i7) matrix(
        [5,0],
        [2,-2]
      );
```

$$(\%o7)\quad \begin{bmatrix} 5 & 0 \\ 2 & -2 \end{bmatrix}$$

DIAGONAL MATRIX

To define a diagonal matrix, go to Algebra → Enter Matrix, choose type: diagonal as shown below:

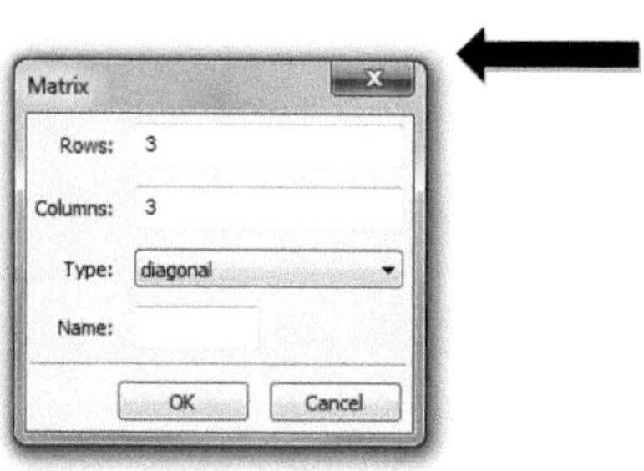

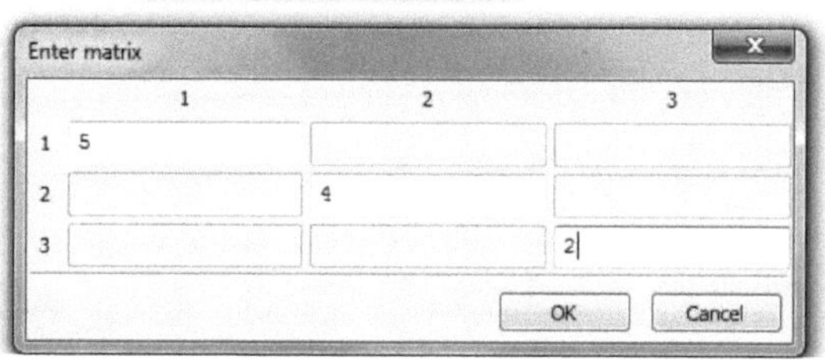

```
(%i8) matrix(
        [5,0,0],
        [0,4,0],
        [0,0,2]
      );
```

$$(\%o8)\quad \begin{bmatrix} 5 & 0 & 0 \\ 0 & 4 & 0 \\ 0 & 0 & 2 \end{bmatrix}$$

SYMMETRIC MATRIX

To define a symmetric matrix, go to Algebra → Enter Matrix, choose type: symmetric as shown below:

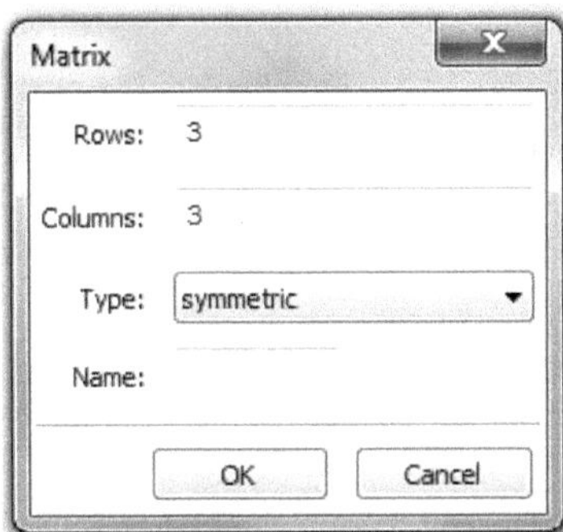

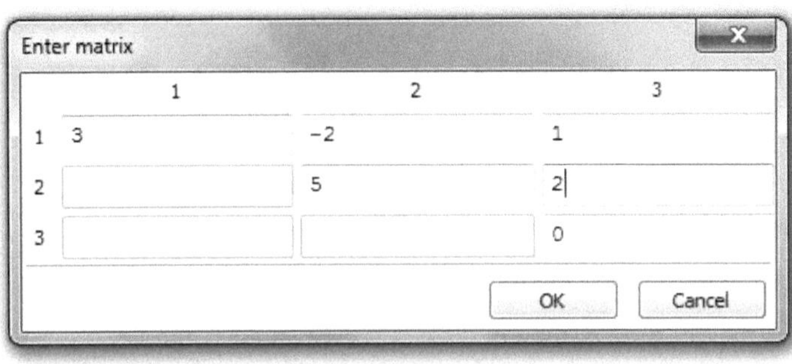

```
(%i9) matrix(
        [3,-2,1],
        [-2,5,2],
        [1,2,0]
      );
```

$$(\%o9)\quad \begin{bmatrix} 3 & -2 & 1 \\ -2 & 5 & 2 \\ 1 & 2 & 0 \end{bmatrix}$$

ANTISYMMETRIC MATRIX

To define an antisymmetric matrix, go to Algebra → Enter Matrix, choose type: antisymmetric as shown below:

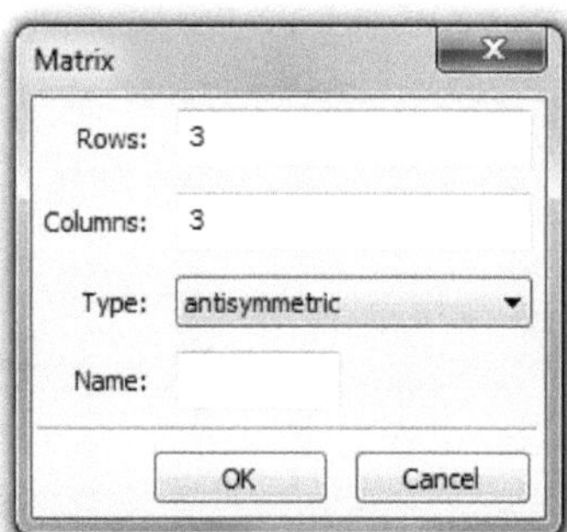

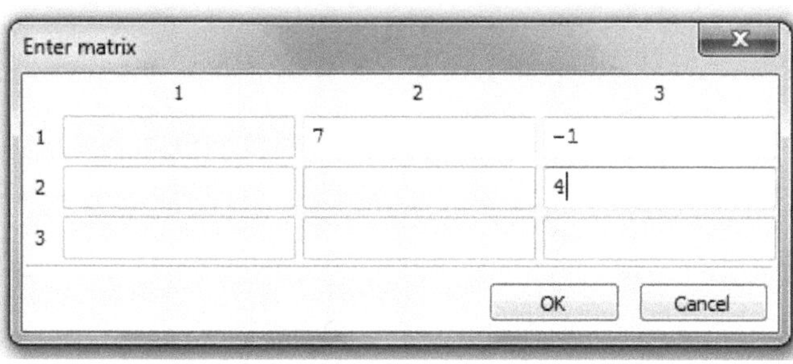

```
(%i10) matrix(
         [0,7,-1],
         [-(7),0,4],
         [-(-1),-(4),0]
       );
```

$$(\%o10)\quad \begin{bmatrix} 0 & 7 & -1 \\ -7 & 0 & 4 \\ 1 & -4 & 0 \end{bmatrix}$$

INVERT MATRIX

When invoked from the *Algebra* menu, the menu item *Invert matrix* produces the inverse of the matrix referred to with %, or of a matrix referred to by name or listen in the *INPUT* line.

Find the inverse matrix of the following matrices

$$\text{a)}\quad \begin{matrix} 1 & 2 & -3 \\ 3 & 2 & 1 \\ 5 & 5 & 3 \end{matrix}$$

Define the matrix

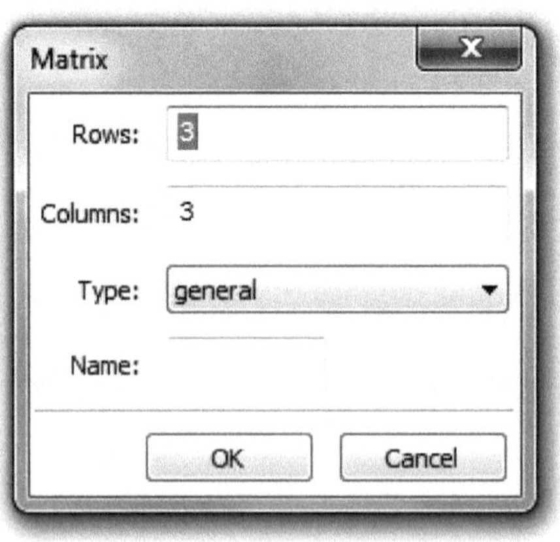

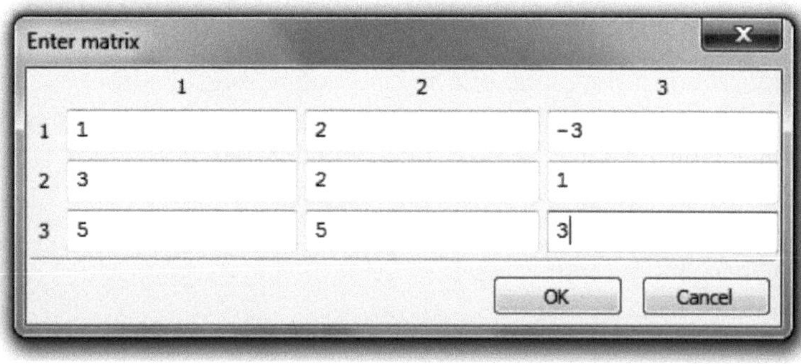


```
(%i11) matrix(
        [1,2,-3],
        [3,2,1],
        [5,5,3]
       );
```

$$(\%o11)\quad \begin{bmatrix} 1 & 2 & -3 \\ 3 & 2 & 1 \\ 5 & 5 & 3 \end{bmatrix}$$

Go to Algebra → Invert Matrix (or type in the input invert %)

```
(%i12) invert(%);
```

$$(\%o12)\quad \begin{bmatrix} -\frac{1}{22} & \frac{21}{22} & -\frac{4}{11} \\ \frac{2}{11} & -\frac{9}{11} & \frac{5}{11} \\ -\frac{5}{22} & -\frac{5}{22} & \frac{2}{11} \end{bmatrix}$$

CHARACTERISTIC POLYNOMIAL

The characteristic polynomial of a square matrix **A** results from expanding the determinant of the matrix **A**-x**I** where **I** is the identity matrix with the same dimensions of **A**, i.e., *charpoly*(**A**) = det(**A**-x**I**).

Find the det($A-xI$) of the following matrices

a) $\begin{bmatrix} 3 & 4 \\ 5 & -1 \end{bmatrix}$

Go to Algebra →Characteristic Polynomial

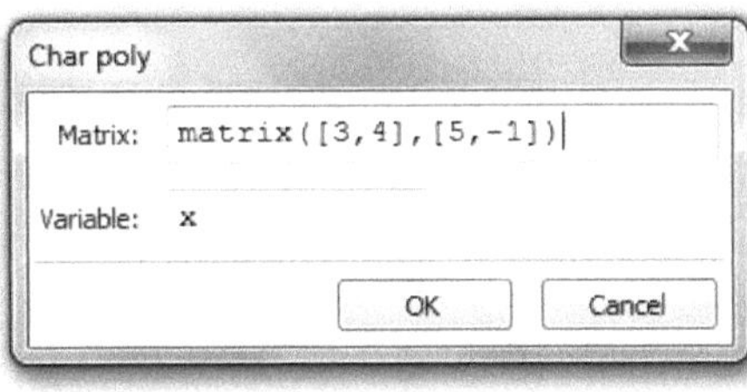


```
(%i13) charpoly(matrix([3,4],[5,-1]), x), expand;
```

(%o13) $x^2-2x-23$

b) $\begin{matrix} 3 & 2 & -1 \\ 1 & 2 & 5 \\ 3 & -2 & 1 \end{matrix}$

Go to Algebra → Characteristic Polynomial

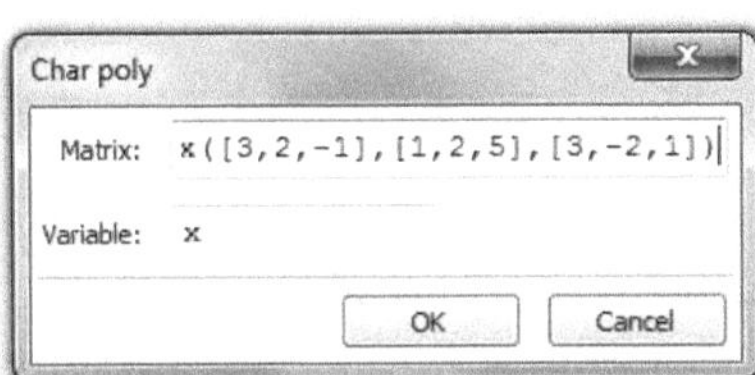


```
(%i14) charpoly(matrix([3,2,-1],[1,2,5],[3,-2,1]), x), expand;
```

(%o14) $-x^3+6x^2-22x+72$

DETERMINANT

The *Algebra* → *Determinant* menu item calculates the determinant of a matrix.

Find the determinant of the following matrices

a) $\begin{matrix} 3 & -1 & 2 \\ -1 & 4 & 2 \\ 2 & 2 & 1 \end{matrix}$

Define the matrix using Maxima first, then go to *Algebra Determinant* or type determinant (%)

```
(%i15) matrix([3,-1,2],
       [-1,4,2],
       [2,2,1]
       );
```

(%o15) $\begin{bmatrix} 3 & -1 & 2 \\ -1 & 4 & 2 \\ 2 & 2 & 1 \end{bmatrix}$

```
(%i16) determinant(%);
(%o16) -25
```

b) $\begin{matrix} 1 & 2 & -3 \\ 3 & 2 & 1 \\ 5 & 5 & 3 \end{matrix}$

Define the matrix using Maxima first, then go to *Algebra Determinant* or type determinant (%)

```
(%i18) matrix(
         [1,2,-3],
         [3,2,1],
         [5,5,3]
       );
```

$$(\%o18)\quad \begin{bmatrix} 1 & 2 & -3 \\ 3 & 2 & 1 \\ 5 & 5 & 3 \end{bmatrix}$$

```
(%i19) determinant (%);
(%o19) -22
```

EIGENVALUES

The *Algebra* → *Eigenvalues* menu item calculates the eigenvalues of a matrix, i.e., it finds the roots of the characteristic polynomial of the matrix. Here is an example of this function applied to matrix **A**:

```
(%i22) A:matrix([3,-1,2],
       [-1,4,2],
       [2,2,1]
       );
```

$$(\%o22)\quad \begin{bmatrix} 3 & -1 & 2 \\ -1 & 4 & 2 \\ 2 & 2 & 1 \end{bmatrix}$$

```
(%i23) eigenvalues(%);
```

$$(\%o23)\quad [[-\frac{\sqrt{29}-3}{2},\frac{\sqrt{29}+3}{2},5],[1,1,1]]$$

The output of function *eigenvalues* consists of two lists. The first list is the list of eigenvalues, and the second list is the multiplicity of those values.

EIGENVECTORS

The *Algebra* → *Eigenvectors* menu item calculates the eigenvectors of a matrix, i.e., it solves for the vectors **v** from the eigenvalue equation $Av = xv$.

For example, for the matrix **A** defined above:

```
(%i27) eigenvectors(A);
```

$$(\%o27)\ [[[-\frac{\sqrt{29}-3}{2},\frac{\sqrt{29}+3}{2},5],[1,1,1]],[[[1,\frac{\sqrt{29}+3}{10},\frac{\sqrt{29}+3}{5}]],[[1,-\frac{\sqrt{29}-3}{10},\frac{\sqrt{29}-3}{5}]],[[0,1,\frac{1}{2}]]]]$$

The output of this command includes the eigenvalues as the first element consisting of two lists as described above. The remaining lists are the eigenvectors of the matrix corresponding to the eigenvalues listed first.

ADJOINT MATRIX

The adjoint matrix produced by the *Algebra* → *Adjoint matrix* menu item corresponds to the definition of the *adjugate* matrix. In the following example, we first put together a complex matrix **A** and then calculate the adjoint or adjugate matrix.

First, we define a couple of 3x3 real matrices *AR* and *AI*:

$$AR = \begin{matrix} 2 & 1 & -1 \\ 3 & 5 & 6 \\ 9 & 2 & 8 \end{matrix} \text{ and } AI = \begin{matrix} 3 & -1 & 2 \\ 7 & 6 & 2 \\ 1 & 2 & 3 \end{matrix}$$

```
(%i1) AR : matrix([2,1,-1],[3,5,6],[9,2,8]);
       [ 2  1  -1 ]
(%o1)  [ 3  5   6 ]
       [ 9  2   8 ]

(%i2) AI : matrix([3,-1,2],[7,6,2],[1,2,3]);
       [ 3  -1  2 ]
(%o2)  [ 7   6  2 ]
       [ 1   2  3 ]
```

Then, we put together matrix $A = AI + i \times AR$:

```
(%i4) A: AI + AR * %i;
       [ 2 %i + 3   %i - 1     2 - %i  ]
(%o4)  [ 3 %i + 7   5 %i + 6   6 %i + 2 ]
       [ 9 %i + 1   2 %i + 2   8 %i + 3 ]
```

Next, we invoke the *Algebra →Adjoint matrix* menu item to produce the adjugate (or adjoint) matrix:

```
(%i5) adjoint(%);
       [ (5 %i + 6) (8 %i + 3) - (2 %i + 2) (6 %i + 2)   (2 - %i) (2 %i + 2) - (%i - 1) (8 %i + 3)   (%i - 1) (6 %i + 2) - (2 - %i) (5 %i + 6) ]
(%o5)  [ (6 %i + 2) (9 %i + 1) - (3 %i + 7) (8 %i + 3)   (2 %i + 3) (8 %i + 3) - (2 - %i) (9 %i + 1)   (2 - %i) (3 %i + 7) - (2 %i + 3) (6 %i + 2) ]
       [ (2 %i + 2) (3 %i + 7) - (5 %i + 6) (9 %i + 1)   (%i - 1) (9 %i + 1) - (2 %i + 2) (2 %i + 3)   (2 %i + 3) (5 %i + 6) - (%i - 1) (3 %i + 7) ]
```

We can use function *rectform* to simplify the matrix to:

```
(%i6) ratsimp(%);
```

$$(\%o6)\quad \begin{bmatrix} 47\,\%i-14 & 7\,\%i+17 & -8\,\%i-25 \\ -41\,\%i-49 & 13\,\%i-18 & 23-23\,\%i \\ 47-39\,\%i & -18\,\%i-12 & 23\,\%i+18 \end{bmatrix}$$

TRANSPOSE MATRIX

The *Algebra* → *Transpose matrix* menu item produces the transpose of a matrix.

For example, for the matrix *A* defined above, we have:

```
(%i8) transpose(A);
```

$$(\%o8)\quad \begin{bmatrix} 2\,\%i+3 & 3\,\%i+7 & 9\,\%i+1 \\ \%i-1 & 5\,\%i+6 & 2\,\%i+2 \\ 2-\%i & 6\,\%i+2 & 8\,\%i+3 \end{bmatrix}$$

APPLICATION

1) Find the inverse matrix for the following matrices

a) $A = \begin{matrix} 1 & 2 & 3 \\ -2 & 3 & 1 \\ 4 & 5 & -2 \end{matrix}$

b) $B = \begin{matrix} 3 & -2 & 1 \\ 5 & 6 & 2 \\ 1 & 0 & -3 \end{matrix}$

c) $C = \begin{matrix} -2 & 3 & 0 \\ 4 & 1 & -3 \\ 2 & 0 & 1 \end{matrix}$

d) $D = \begin{matrix} 2 & 1 & 3 \\ -1 & 2 & 0 \\ 3 & 2 & 1 \end{matrix}$

e) $E = \begin{matrix} 6 & 2 & 8 \\ -3 & 4 & 1 \\ 4 & -4 & 5 \end{matrix}$

f) $F = \begin{matrix} 3 & 2 \\ -3 & 4 \end{matrix}$

g) $G = \begin{matrix} 4 & 2 & 2 \\ 0 & 1 & 2 \\ 1 & 0 & 3 \end{matrix}$

h) $H = \begin{matrix} 4 & 0 & 0 \\ 0 & -3 & 0 \\ 0 & 0 & 2 \end{matrix}$

2) Find the det $(A - xI)$ of the following matrices

a) $A = \begin{matrix} 1 & 2 & 3 \\ -2 & 3 & 1 \\ 4 & 5 & -2 \end{matrix}$

b) $B = \begin{matrix} 3 & -2 & 1 \\ 5 & 6 & 2 \\ 1 & 0 & -3 \end{matrix}$

c) $C = \begin{matrix} -2 & 3 & 0 \\ 4 & 1 & -3 \\ 2 & 0 & 1 \end{matrix}$

d) $$D = \begin{matrix} 2 & 1 & 3 \\ -1 & 2 & 0 \\ 3 & 2 & 1 \end{matrix}$$

e) $$E = \begin{matrix} 6 & 2 & 8 \\ -3 & 4 & 1 \\ 4 & -4 & 5 \end{matrix}$$

f) $$F = \begin{matrix} 3 & 2 \\ -3 & 4 \end{matrix}$$

g) $$G = \begin{matrix} 4 & 2 & 2 \\ 0 & 1 & 2 \\ 1 & 0 & 3 \end{matrix}$$

h) $$H = \begin{matrix} 4 & 0 & 0 \\ 0 & -3 & 0 \\ 0 & 0 & 2 \end{matrix}$$

3) Find the determinant of the following matrices

a) $$A = \begin{matrix} 1 & 2 & 3 \\ -2 & 3 & 1 \\ 4 & 5 & -2 \end{matrix}$$

b) $$B = \begin{matrix} 3 & -2 & 1 \\ 5 & 6 & 2 \\ 1 & 0 & -3 \end{matrix}$$

c) $$C = \begin{matrix} -2 & 3 & 0 \\ 4 & 1 & -3 \\ 2 & 0 & 1 \end{matrix}$$

d) $$D = \begin{matrix} 2 & 1 & 3 \\ -1 & 2 & 0 \\ 3 & 2 & 1 \end{matrix}$$

e) $E = \begin{matrix} 6 & 2 & 8 \\ -3 & 4 & 1 \\ 4 & -4 & 5 \end{matrix}$

f) $F = \begin{matrix} 3 & 2 \\ -3 & 4 \end{matrix}$

g) $G = \begin{matrix} 4 & 2 & 2 \\ 0 & 1 & 2 \\ 1 & 0 & 3 \end{matrix}$

h) $H = \begin{matrix} 4 & 0 & 0 \\ 0 & -3 & 0 \\ 0 & 0 & 2 \end{matrix}$

4) Find the adjoint matrix

a) $A = \begin{matrix} 1 & 2 & 3 \\ -2 & 3 & 1 \\ 4 & 5 & -2 \end{matrix}$

b) $B = \begin{matrix} 3 & -2 & 1 \\ 5 & 6 & 2 \\ 1 & 0 & -3 \end{matrix}$

c) $C = \begin{matrix} -2 & 3 & 0 \\ 4 & 1 & -3 \\ 2 & 0 & 1 \end{matrix}$

d) $D = \begin{matrix} 2 & 1 & 3 \\ -1 & 2 & 0 \\ 3 & 2 & 1 \end{matrix}$

e) $E = \begin{matrix} 6 & 2 & 8 \\ -3 & 4 & 1 \\ 4 & -4 & 5 \end{matrix}$

f) $F = \begin{matrix} 3 & 2 \\ -3 & 4 \end{matrix}$

g) $G = \begin{matrix} 4 & 2 & 2 \\ 0 & 1 & 2 \\ 1 & 0 & 3 \end{matrix}$

h) $H = \begin{matrix} 4 & 0 & 0 \\ 0 & -3 & 0 \\ 0 & 0 & 2 \end{matrix}$

5) Find the transpose matrix of the following matrices

i) $A = \begin{matrix} 1 & 2 & 3 \\ -2 & 3 & 1 \\ 4 & 5 & -2 \end{matrix}$

j) $B = \begin{matrix} 3 & -2 & 1 \\ 5 & 6 & 2 \\ 1 & 0 & -3 \end{matrix}$

k) $C = \begin{matrix} -2 & 3 & 0 \\ 4 & 1 & -3 \\ 2 & 0 & 1 \end{matrix}$

l) $D = \begin{matrix} 2 & 1 & 3 \\ -1 & 2 & 0 \\ 3 & 2 & 1 \end{matrix}$

m) $E = \begin{matrix} 6 & 2 & 8 \\ -3 & 4 & 1 \\ 4 & -4 & 5 \end{matrix}$

n) $F = \begin{matrix} 3 & 2 \\ -3 & 4 \end{matrix}$

o) $G = \begin{matrix} 4 & 2 & 2 \\ 0 & 1 & 2 \\ 1 & 0 & 3 \end{matrix}$

p) $H = \begin{matrix} 4 & 0 & 0 \\ 0 & -3 & 0 \\ 0 & 0 & 2 \end{matrix}$

4.3 FUNCTIONS FOR CREATING MATRICES

We present in this section examples of *Maxima* functions to generate matrices. Some functions for creating matrices that are available in the *Algebra* menu were introduced above (*genmatrix, transpose*). The following examples demonstrate the use of additional functions:

COPYMATRIX

Use *copymatrix* to copy a matrix into a variable name.

For example, first create matrix $A = \begin{matrix} 1 & 2 & 4 \\ 3 & 4 & 2 \\ 2 & -1 & 1 \end{matrix}$

```
(%i1) A: matrix(
        [1,2,4],
        [3,4,2],
        [2,-1,1]
      );
```

$$(\%o1) \quad \begin{bmatrix} 1 & 2 & 4 \\ 3 & 4 & 2 \\ 2 & -1 & 1 \end{bmatrix}$$

Then, copy matrix A into B using *copymatrix*:

```
(%i2) B: copymatrix(A);
```

$$(\%o2) \quad \begin{bmatrix} 1 & 2 & 4 \\ 3 & 4 & 2 \\ 2 & -1 & 1 \end{bmatrix}$$

COLUMNVECTOR

A column vector is a matrix of n rows and 1 column. Function *columnvector* lets you build a column vector out of a list of values, e.g.: $\begin{matrix} 3 \\ 4 \\ -2 \end{matrix}$

```
(%i6) v:columnvector([3,4,-2]);
(%o6) columnvector([3,4,-2])
```

DIAG

Function *diag*, which needs to be loaded separately, allows you to build a diagonal matrix based on two or more matrices.

For example, using matrices *A* and *B*, defined above, we can build the following diagonal matrix:

```
(%i7) load("diag");
(%o7) C:/PROGRA~1/MAXIMA~1.2/share/maxima/5.31.2/share/contrib/diag.mac

(%i8) diag([A,B]);
```

$$(\%o8) \quad \begin{bmatrix} 1 & 2 & 4 & 0 & 0 & 0 \\ 3 & 4 & 2 & 0 & 0 & 0 \\ 2 & -1 & 1 & 0 & 0 & 0 \\ 0 & 0 & 0 & 1 & 2 & 4 \\ 0 & 0 & 0 & 3 & 4 & 2 \\ 0 & 0 & 0 & 2 & -1 & 1 \end{bmatrix}$$

DIAGMATRIX

Function *diagmatrix* (*n,a*) creates a diagonal matrix of dimensions $n \times n$ with all its diagonal elements equal to *a*:

```
(%i9) diagmatrix(4,2);
```

$$(\%o9) \begin{bmatrix} 2 & 0 & 0 & 0 \\ 0 & 2 & 0 & 0 \\ 0 & 0 & 2 & 0 \\ 0 & 0 & 0 & 2 \end{bmatrix}$$

Function *diagmatrix* can be used to generate an identity matrix

```
(%i10) diagmatrix(5,1);
```

$$(\%o10) \begin{bmatrix} 1 & 0 & 0 & 0 & 0 \\ 0 & 1 & 0 & 0 & 0 \\ 0 & 0 & 1 & 0 & 0 \\ 0 & 0 & 0 & 1 & 0 \\ 0 & 0 & 0 & 0 & 1 \end{bmatrix}$$

EMATRIX

Function *ematrix* (*n,m,a,i,j*) creates a matrix of dimensions $n \times m$ full of zero elements except for element [*i,j*] which is replaced by the value *a*, *e.g.*,

```
(%i11) ematrix(4,4,3,2,1);
```

$$(\%o11) \begin{bmatrix} 0 & 0 & 0 & 0 \\ 3 & 0 & 0 & 0 \\ 0 & 0 & 0 & 0 \\ 0 & 0 & 0 & 0 \end{bmatrix}$$

ENTERMATRIX

Function *entermatrix* (*n,m*) provides for an interactive, if long, way to enter a matrix

```
(%i12) entermatrix(3,4);
Row 1 Column 1:  3;
Row 1 Column 2:  -2;
Row 1 Column 3:  5;
Row 1 Column 4:  2;
Row 2 Column 1:  1;
Row 2 Column 2:  9;
Row 2 Column 3:  4;
Row 2 Column 4:  -9;
Row 3 Column 1:  4;
Row 3 Column 2:  6;
Row 3 Column 3:  8;
Row 3 Column 4:  -2;
Matrix entered.
```

$$(\%o12)\quad \begin{bmatrix} 3 & -2 & 5 & 2 \\ 1 & 9 & 4 & -9 \\ 4 & 6 & 8 & -2 \end{bmatrix}$$

IDENT

Function *ident*(*n*) allows to create an $n \times n$ identity matrix

$$I_3 = \begin{matrix} 1 & 0 & 0 \\ 0 & 1 & 0 \\ 0 & 0 & 1 \end{matrix}$$

```
(%i14) I3: ident(3);
```

$$(\%o14)\quad \begin{bmatrix} 1 & 0 & 0 \\ 0 & 1 & 0 \\ 0 & 0 & 1 \end{bmatrix}$$

```
(%i13) I4: ident(4);
```

$$(\%o13)\quad \begin{bmatrix} 1 & 0 & 0 & 0 \\ 0 & 1 & 0 & 0 \\ 0 & 0 & 1 & 0 \\ 0 & 0 & 0 & 1 \end{bmatrix}$$

ZEROMATRIX

Function *zeromatrix* (*m,n*) creates a matrix of dimensions $m \times n$ such that all its elements are zero values.

```
(%i15) zeromatrix(2,3);
```

$$(\%o15)\quad \begin{bmatrix} 0 & 0 & 0 \\ 0 & 0 & 0 \end{bmatrix}$$

SUBMATRIX

Function *submatrix* allows the user to extract a submatrix out of a matrix.

To illustrate the use of function *submatrix* considers the 4×4 matrix A:

```
(%i16) A: matrix(
         [3,2,-1,2],
         [5,7,8,-6],
         [0,4,8,3],
         [-2,-1,4,3]
       );
```

$$(\%o16) \quad \begin{bmatrix} 3 & 2 & -1 & 2 \\ 5 & 7 & 8 & -6 \\ 0 & 4 & 8 & 3 \\ -2 & -1 & 4 & 3 \end{bmatrix}$$

To extract a matrix by eliminating rows from *i1* to *i2* and columns from *j1* to *j2* out of the matrix

A, use the general call *submatrix(i1,i2,A,j1,j2)*. In the following example, we eliminate rows

1 and 3 and columns 2 to 3 out of matrix *A* and store the resulting matrix into *B*:

```
(%i17) submatrix(2,3,A,2,3);
```

$$(\%o17) \quad \begin{bmatrix} 3 & 2 \\ -2 & 3 \end{bmatrix}$$

To eliminate rows from *i1* to *i2*, only, use the modified call *submatrix(i1,i2,A)*

```
(%i18) submatrix(2,3,A);
```

$$(\%o18) \quad \begin{bmatrix} 3 & 2 & -1 & 2 \\ -2 & -1 & 4 & 3 \end{bmatrix}$$

To eliminate columns from *j1* to *j2*, only, use the modified call *submatrix(A,j1,j2)*

```
(%i19) submatrix(A,2,3);
        [  3   2 ]
        [        ]
        [  5  -6 ]
(%o19)  [        ]
        [  0   3 ]
        [        ]
        [ -2   3 ]
```

4.4 FUNCTIONS FOR MANIPULATING MATRICES

The following functions allow the user to add or extract rows and columns out of matrices.

To illustrate the use of these functions we will refer to matrix *A* defined above and repeated here:

```
(%i20) A;
        [  3   2  -1   2 ]
        [                ]
        [  5   7   8  -6 ]
(%o20)  [                ]
        [  0   4   8   3 ]
        [                ]
        [ -2  -1   4   3 ]
```

COL

Function *col* extracts a column out of a matrix

```
(%i21) C2: col(A,2);
(%o21) [ 2]
       [ 7]
       [ 4]
       [-1]

(%i22) C3: col(A,3);
(%o22) [-1]
       [ 8]
       [ 8]
       [ 4]
```

ROW

Function *row* extracts a row out of a matrix

```
(%i24) R2: row(A,2);
(%o24) [5 7 8 -6]

(%i25) R3: row(A,3);
(%o25) [0 4 8 3]
```

ADDCOL

Function *addcol* is used to append one or more columns to a matrix.

We need to add the following column [5, 4, 3, 2] to the matrix A

```
(%i26) A5C: addcol(A,[5,4,3,2]);
```

$$(\%o26)\quad \begin{bmatrix} 3 & 2 & -1 & 2 & 5 \\ 5 & 7 & 8 & -6 & 4 \\ 0 & 4 & 8 & 3 & 3 \\ -2 & -1 & 4 & 3 & 2 \end{bmatrix}$$

We need now to add the following two columns [3, 2, 1, 4] and [4, 3, 2, 1] to A

```
(%i27) A6C: addcol(A,[3,2,1,4],[4,3,2,1]);
```

$$(\%o27)\quad \begin{bmatrix} 3 & 2 & -1 & 2 & 3 & 4 \\ 5 & 7 & 8 & -6 & 2 & 3 \\ 0 & 4 & 8 & 3 & 1 & 2 \\ -2 & -1 & 4 & 3 & 4 & 1 \end{bmatrix}$$

ADDROW

Function *addrow* is used to append one or more rows to a matrix.

We need to add the following row [5, 4, 3, 2] to the matrix A

```
(%i29) A5R: addrow(A,[5,4,3,2]);
```

$$(\%o29)\quad \begin{bmatrix} 3 & 2 & -1 & 2 \\ 5 & 7 & 8 & -6 \\ 0 & 4 & 8 & 3 \\ -2 & -1 & 4 & 3 \\ 5 & 4 & 3 & 2 \end{bmatrix}$$

We need now to add the following two rows [3, 2, 1, 4] and [4, 3, 2, 1] to A

```
(%i30) A6R: addrow(A,[3, 2, 1, 4] ,[4, 3, 2, 1] );
```

$$(\%o30)\ \begin{bmatrix} 3 & 2 & -1 & 2 \\ 5 & 7 & 8 & -6 \\ 0 & 4 & 8 & 3 \\ -2 & -1 & 4 & 3 \\ 3 & 2 & 1 & 4 \\ 4 & 3 & 2 & 1 \end{bmatrix}$$

4.5 MATRIX OPERATIONS

MATRIX SIZE

The size of a matrix can be determined by function *matrix_size*

```
(%i7) A: matrix(
       [1,2,3],
       [3,5,-2]
      );
```

$$(\%o7)\ \begin{bmatrix} 1 & 2 & 3 \\ 3 & 5 & -2 \end{bmatrix}$$

```
(%i9) sA : matrix_size(A);
(%o9) [2,3]
```

The number of rows and columns can be extracted by using:

```
(%i11) nrows: sA[1];
(%o11) 2
```

```
(%i12) ncols: sA[2]
(%o12) 3
```

Basic matrix operations include addition, subtraction, multiplication, division, and powers.

To illustrate those operations we will use the following matrices A and B

```
(%i1) A: matrix(
        [-2,3,4],
        [5,-6,8],
        [2,1,-1]
      );
```

$$(\%o1)\quad \begin{bmatrix} -2 & 3 & 4 \\ 5 & -6 & 8 \\ 2 & 1 & -1 \end{bmatrix}$$

```
(%i2) B: matrix(
        [2,5,7],
        [-2,3,4],
        [6,-2,4]
      );
```

$$(\%o2)\quad \begin{bmatrix} 2 & 5 & 7 \\ -2 & 3 & 4 \\ 6 & -2 & 4 \end{bmatrix}$$

ADDITION AND SUBTRACTION

Addition and subtraction are term-by-term operations on matrices of the same dimensions.

The examples below include linear combinations of additions and subtractions

- $A + B$

```
(%i3) A + B;
```

$$(\%o3)\quad \begin{bmatrix} 0 & 8 & 11 \\ 3 & -3 & 12 \\ 8 & -1 & 3 \end{bmatrix}$$

- $A - B$

```
(%i4) A - B;
```

$$(\%o4) \begin{bmatrix} -4 & -2 & -3 \\ 7 & -9 & 4 \\ -4 & 3 & -5 \end{bmatrix}$$

MULTIPLICATION

Multiplication can be *term-by-term*, in which case we use an asterisk for the multiplication symbol

```
(%i5) A * B;
```

$$(\%o5) \begin{bmatrix} -4 & 15 & 28 \\ -10 & -18 & 32 \\ 12 & -2 & -4 \end{bmatrix}$$

Traditional, *non-commutative, matrix multiplication* is achieved by using a dot (.) as the multiplication symbol

```
(%i6) A . B;
```

$$(\%o6) \begin{bmatrix} 14 & -9 & 14 \\ 70 & -9 & 43 \\ -4 & 15 & 14 \end{bmatrix}$$

```
(%i7) B . A;
```

$$(\%o7) \begin{bmatrix} 35 & -17 & 41 \\ 27 & -20 & 12 \\ -14 & 34 & 4 \end{bmatrix}$$

POWER

A matrix raised to a scalar exponential produces a term-by-term exponentiation

```
(%i8) A^2;
```

$$(\%o8)\quad \begin{bmatrix} 4 & 9 & 16 \\ 25 & 36 & 64 \\ 4 & 1 & 1 \end{bmatrix}$$

```
(%i9) B^3;
```

$$(\%o9)\quad \begin{bmatrix} 8 & 125 & 343 \\ -8 & 27 & 64 \\ 216 & -8 & 64 \end{bmatrix}$$

A scalar base raised to a matrix exponent is also a term-by-term operation

```
(%i10) exp (A);
```

$$(\%o10)\quad \begin{bmatrix} \%e^{-2} & \%e^{3} & \%e^{4} \\ \%e^{5} & \%e^{-6} & \%e^{8} \\ \%e^{2} & \%e & \%e^{-1} \end{bmatrix}$$

```
(%i11) 4^B;
```

$$(\%o11)\quad \begin{bmatrix} 16 & 1024 & 16384 \\ \frac{1}{16} & 64 & 256 \\ 4096 & \frac{1}{16} & 256 \end{bmatrix}$$

Matrix exponentiation uses a double caret (^^) and represents the result of repeated matrix multiplication, i.e., A^^2 = A.A, A^^3 = A^^2.A, and so on:

```
(%i12) A^^2;
```

$$(\%o12)\quad \begin{bmatrix} 27 & -20 & 12 \\ -24 & 59 & -36 \\ -1 & -1 & 17 \end{bmatrix}$$

```
(%i13) A^^3;
```

$$(\%o13)\quad \begin{bmatrix} -130 & 213 & -64 \\ 271 & -462 & 412 \\ 31 & 20 & -29 \end{bmatrix}$$

```
(%i14) B^^2;
```

$$(\%o14)\quad \begin{bmatrix} 36 & 11 & 62 \\ 14 & -9 & 14 \\ 40 & 16 & 50 \end{bmatrix}$$

```
(%i15) B^^4;
```

$$(\%o15)\quad \begin{bmatrix} 3930 & 1289 & 5486 \\ 938 & 459 & 1442 \\ 3664 & 1096 & 5204 \end{bmatrix}$$

DIVISION

Division of matrices is a term-by-term operation

```
(%i16) A/B;
```

$$(\%o16)\quad \begin{bmatrix} -1 & \frac{3}{5} & \frac{4}{7} \\ -\frac{5}{2} & -2 & 2 \\ \frac{1}{3} & -\frac{1}{2} & -\frac{1}{4} \end{bmatrix}$$

```
(%i17) B/A;
```

$$(\%o17)\quad \begin{bmatrix} -1 & \frac{5}{3} & \frac{7}{4} \\ -\frac{2}{5} & -\frac{1}{2} & \frac{1}{2} \\ 3 & -2 & -4 \end{bmatrix}$$

CONJUGATE

The *conjugate* function used to calculate the complex conjugate of a number can be used to find the conjugate matrix of a matrix of complex numbers

```
(%i18) Z: A + B *%i;
```

$$(\%o18)\quad \begin{bmatrix} 2\,\%i-2 & 5\,\%i+3 & 7\,\%i+4 \\ 5-2\,\%i & 3\,\%i-6 & 4\,\%i+8 \\ 6\,\%i+2 & 1-2\,\%i & 4\,\%i-1 \end{bmatrix}$$

```
(%i19) conjugate(Z);
```

$$(\%o19)\quad \begin{bmatrix} -2\%i-2 & 3-5\%i & 4-7\%i \\ 2\%i+5 & -3\%i-6 & 8-4\%i \\ 2-6\%i & 2\%i+1 & -4\%i-1 \end{bmatrix}$$

APPLICATION

Perform the indicated operations and find the size of each matrix:

1. $(A)+(B)$ Given: $A=\begin{pmatrix} 4 & -3 & 6 \\ -8 & 5 & -9 \end{pmatrix}$ $B=\begin{pmatrix} -5 & 6 & -2 \\ 3 & 7 & -4 \end{pmatrix}$

   $(A)-(B)$ Given: $A=\begin{pmatrix} 4 & -3 & 6 \\ -8 & 5 & -9 \end{pmatrix}$ $B=\begin{pmatrix} -5 & 6 & -2 \\ 3 & 7 & -4 \end{pmatrix}$

   $(A)\times(B)$ Given: $A=\begin{pmatrix} 4 & -3 & 6 \\ -8 & 5 & -9 \end{pmatrix}$ $B=\begin{pmatrix} -5 & 6 & -2 \\ 3 & 7 & -4 \end{pmatrix}$

   $(A)\div(B)$ Given: $A=\begin{pmatrix} 4 & -3 & 6 \\ -8 & 5 & -9 \end{pmatrix}$ $B=\begin{pmatrix} -5 & 6 & -2 \\ 3 & 7 & -4 \end{pmatrix}$

2. $(A)-(B)$ Given: $A=\begin{pmatrix} 6 & -7 \\ -4 & 5 \\ -3 & 2 \end{pmatrix}$ $B=\begin{pmatrix} -8 & 3 \\ 3 & -1 \\ 2 & -8 \end{pmatrix}$

3. $5\begin{pmatrix} -4 & 3 \\ 6 & -2 \end{pmatrix}$

4. $(A)\,(B)$ Given: $A=\begin{pmatrix} 6 & -2 & 3 \\ -4 & 2 & 5 \end{pmatrix}$ $B=\begin{pmatrix} 2 & -3 \\ 4 & -5 \\ 1 & -6 \end{pmatrix}$

5. Let $A = \begin{bmatrix} 2 & 1 \\ 3 & 0 \\ -1 & 4 \end{bmatrix}_{3\times 2}$ and $B = \begin{bmatrix} 2 & 3 & -1 \\ 1 & 0 & 4 \end{bmatrix}_{2\times 3}$

Find AB and BA

6. Let $A = \begin{pmatrix} 1 & 2 & -3 \\ -1 & 0 & 2 \end{pmatrix}$ $B = \begin{pmatrix} 2 & 4 & 0 \\ 3 & -1 & 1 \end{pmatrix}$ $C = \begin{pmatrix} 2 & 1 \\ 1 & 0 \\ -1 & 1 \end{pmatrix}$ and $D = \begin{pmatrix} 1 \\ 2 \\ 0 \end{pmatrix}$

Evaluate the following:

(a) $(A + 2B)\,C$

(b) (AC^2)

4.6 FUNCTIONS FOR LINEAR ALGEBRA OPERATIONS

The functions in this section are used in linear algebra applications. Some functions, such as *adjoint, charpoly, determinant, eigen,* and *invert,* were introduced as part of the *Algebra* menu items.

COEFMATRIX

Function *coefmatrix*([*list of linear equations*],[*list of variables*]) can be used to extract the coefficients from a *list of linear equations* containing the variables in the *list of variables*,

```
(%i1) Eq1: 2*x+5*y+3*z = 25;
(%o1) 3 z+5 y+2 x=25

(%i2) Eq2: 3*x-y+5*z = 105;
(%o2) 5 z-y+3 x=105

(%i3) Eq3: x+y+z = 18;
(%o3) z+y+x=18

(%i4) coefmatrix([Eq1,Eq2,Eq3],[x,y,z]);
```

(%o4) $\begin{bmatrix} 2 & 5 & 3 \\ 3 & -1 & 5 \\ 1 & 1 & 1 \end{bmatrix}$

AUGCOEFMATRIX

Function *augcoefmatrix*([*list of linear equations*],[*list of variables*]) produces an augmented matrix of coefficients similar to that produced by *coefmatrix*, except that the last column contains the negatives of the right-hand side elements corresponding to the *list of equations*. For example, for the system of linear equations used in the *coefmatrix* example shown above, the resulting augmented matrix of coefficients is calculated as follows:

```
(%i5) AA: augcoefmatrix([Eq1,Eq2,Eq3],[x,y,z]);
```

$$(\%o5)\quad \begin{bmatrix} 2 & 5 & 3 & -25 \\ 3 & -1 & 5 & -105 \\ 1 & 1 & 1 & -18 \end{bmatrix}$$

ECHELON AND TRIANGULARIZE

Both functions *echelon*(*A*) and *triangularize*(*A*) produce upper triangular matrices representing row-reduced echelon forms of matrix *A*. The difference between these two functions is that function *echelon* produces a matrix such that its main diagonal elements are reduced to the number *1*. The following examples, using matrices *A* and *AA* created above, illustrate the application of these two functions highlighting their differences:

```
(%i8) echelon(A);
```

$$(\%o8)\quad \begin{bmatrix} 1 & \frac{5}{2} & \frac{3}{2} \\ 0 & 1 & -\frac{1}{17} \\ 0 & 0 & 1 \end{bmatrix}$$

```
(%i9) triangularize(A);
```

$$(\%o9)\quad \begin{bmatrix} 2 & 5 & 3 \\ 0 & -17 & 1 \\ 0 & 0 & 10 \end{bmatrix}$$

```
(%i10) echelon(AA);
```

$$(\%o10)\quad \begin{bmatrix} 1 & \frac{5}{2} & \frac{3}{2} & -\frac{25}{2} \\ 0 & 1 & -\frac{1}{17} & \frac{135}{17} \\ 0 & 0 & 1 & -\frac{109}{10} \end{bmatrix}$$

```
(%i11) triangularize(AA);
```

$$(\%o11)\quad \begin{bmatrix} 2 & 5 & 3 & -25 \\ 0 & -17 & 1 & -135 \\ 0 & 0 & 10 & -109 \end{bmatrix}$$

MATTRACE

The *mattrace,* the function is available by loading package "nchrpl".

Function *mattrace* calculates the trace of a matrix (i.e., the sum of its main diagonal elements).

Find the trace of the matrices A and AA defined above

```
(%i12) load("nchrpl");
(%o12) C:/PROGRA~1/MAXIMA~1.2/share/maxima/5.31.2/share/matrix/nchrpl.mac

(%i13) mattrace(A);
(%o13) 2

(%i16) mattrace(AA);
(%o16) 2
```

MINOR

The minor matrix (i,j) of a matrix A is the matrix that results from eliminating row i and column j. Minor matrices are used, for example, in the calculation of determinants.

```
(%i18) A;
```

$$(\%o18) \begin{bmatrix} 2 & 5 & 3 \\ 3 & -1 & 5 \\ 1 & 1 & 1 \end{bmatrix}$$

```
(%i19) minor(A,2,2);
```

$$(\%o19) \begin{bmatrix} 2 & 3 \\ 1 & 1 \end{bmatrix}$$

```
(%i20) minor(A,1,2);
```

$$(\%o20) \begin{bmatrix} 3 & 5 \\ 1 & 1 \end{bmatrix}$$

```
(%i21) minor(A,3,3);
```

$$(\%o21) \begin{bmatrix} 2 & 5 \\ 3 & -1 \end{bmatrix}$$

PERMANENT

Function *permanent* calculates the *permanent* of a matrix.

The permanent is an analog of a determinant where all the signs in the expansion by minors are taken as positive. The permanent of a matrix A is the coefficient of $x_1 \ldots x_n$ in

$$\prod_{i=1}^{n}\left(a_i 1x_1 + a_{i2}x_2 + \ldots + a_{in}x_n\right) \tag{4.1}$$

Another equation is the Ryser formula

$$\mathrm{perm}(\mathrm{a}_{ij}) = (-1)^n \sum_{s\subseteq\{1\ldots n\}} (-1)^{|s|} \prod_{i=1}^{n} \sum_{j\in s} a_{ij} \tag{4.2}$$

where the sum is over all subsets of $\{1,0 = \ldots, n\}$, and $|s|$ is the number of elements in s. Muir uses the notation $\big|_+ \; \big|_+$ to denote a permanent.

```
(%i22) permanent(A);
(%o22)/R/ 54
```

RANK

Function *rank* calculates the *rank* of a matrix.

The *rank* of a matrix A is the dimension of the vector space generated (or spanned) by its columns. This is the same as the dimension of the space spanned by its rows. There are multiple equivalent definitions of rank. A matrix's rank is one of its most fundamental characteristics.

The rank is commonly denoted rank(A) or rk(A); sometimes the parentheses are not written, as in rank A.

To find the rank of a matrix, we simply transform the matrix to its row echelon form and count the number of non-zero rows.

```
(%i25) A;
        [2   5  3]
(%o25)  [3  -1  5]
        [1   1  1]
(%i26) rank(A);
(%o26) 3
```

```
(%i27) AA;
        [2   5  3   -25]
(%o27)  [3  -1  5  -105]
        [1   1  1   -18]
(%i28) rank(AA);
(%o28) 3
```

TRACEMATRIX

Function *tracematrix*, which needs to be loaded with package *functs*, calculates the trace of a matrix.

```
(%i25) A;
        [2   5  3]
(%o25)  [3  -1  5]
        [1   1  1]
```

```
(%i29) load("functs");
(%o29) C:/PROGRA~1/MAXIMA~1.2/share/maxima/5.31.2/share/simplification/functs.mac

(%i30) tracematrix(A);
(%o30) 2
```

4.7 FUNCTIONS FOR MATRIX DECOMPOSITION

Matrix decomposition is useful in linear algebra applications. *Maxima* provides the following functions for matrix decomposition:

CHOLESKY

Function *Cholesky* produces the Cholesky decomposition of a symmetric, positive definite matrix.

The Cholesky decomposition or Cholesky factorization is a decomposition of a Hermitian, positive-definite matrix into the product of a lower triangular matrix and its conjugate transpose, which is useful e.g. for efficient numerical solutions and Monte Carlo simulations.

It was discovered by André-Louis Cholesky for real matrices.

When it is applicable, the Cholesky decomposition is roughly twice as efficient as the LU decomposition for solving systems of linear equations.

An example is shown next, in which we first define a matrix A:

```
(%i1) A: matrix(
        [5,4,2],
        [4,10,6],
        [2,6,15]
      );
```

$$(\%o1)\quad \begin{bmatrix} 5 & 4 & 2 \\ 4 & 10 & 6 \\ 2 & 6 & 15 \end{bmatrix}$$

The Cholesky decomposition results in

```
(%i2) cholesky(A);
```

$$(\%o2)\quad \begin{bmatrix} \sqrt{5} & 0 & 0 \\ \frac{4}{\sqrt{5}} & \frac{\sqrt{34}}{\sqrt{5}} & 0 \\ \frac{2}{\sqrt{5}} & \frac{22}{\sqrt{5}\sqrt{34}} & \frac{\sqrt{193}}{\sqrt{17}} \end{bmatrix}$$

EIGENS_BY_JACOBI

Function *eigens_by_jacobi* calculates the eigenvalues of a symmetric real matrix by the method of Jacobi rotations.

A matrix used in the Jacobi transformation method of diagonalizing matrices. The Jacobi rotation matrix P_{pq} contains 1s along the diagonal, except for the two elements cos ϕ in rows and columns p and q In addition, all off-diagonal elements are zero except the elements sin ϕ and – sin ϕ. The rotation angle ϕ for an initial matrix A is chosen such that

$$\cot(2\phi) = \frac{a_{qq} - a_{pp}}{2a_{pq}}$$

Then the corresponding Jacobi rotation matrix which annihilates the off-diagonal element a_{pq} is

$$\mathbf{P}_{pq} \equiv \begin{bmatrix} 1 & & & & & & 0 \\ & \ddots & & \vdots & & ⋰ & \\ & & \cos\phi & \cdots & 0 & \cdots & \sin\phi & & \\ & \cdots & 0 & \cdots & 1 & \cdots & 0 & \cdots \\ & & -\sin\phi & \cdots & 0 & \cdots & \cos\phi & & \\ & ⋰ & & \vdots & & \ddots & \\ 0 & & & & & & 1 \end{bmatrix}$$

Consider the following example in which we first define a symmetric matrix *A*

```
(%i3) A: matrix(
        [5,4,-2],
        [4,3,4],
        [-2,4,6]
      );
```

$$(\%o3) \quad \begin{bmatrix} 5 & 4 & -2 \\ 4 & 3 & 4 \\ -2 & 4 & 6 \end{bmatrix}$$

The eigenvalues and eigenvectors are calculated as follows:

```
(%i4) eigens_by_jacobi (A);
(%o4) [[9.000000000000002,-2.424428900898052,7.424428900898052],
```

$$\begin{bmatrix} 0.33333333333333 & -0.5137494131693 & -0.79053806319309 \\ 0.66666666666667 & 0.72134090068715 & -0.1876775440787 \\ 0.66666666666667 & -0.4644661941025 & 0.58294657567525 \end{bmatrix}]$$

LU_FACTOR

Function *lu_factor* produces the *LU* decomposition of a matrix.

LU decomposition (where 'LU' stands for 'lower upper', and also called LU factorization) factors a matrix as the product of a lower triangular matrix and an upper triangular matrix. The product sometimes includes a permutation matrix as well. The LU decomposition can be viewed as the matrix form of Gaussian elimination. Computers usually solve square systems of linear equations using the LU decomposition, and it is also a key step when inverting a matrix, or computing the determinant of a matrix.

The example shown below uses matrix *A* defined above:

```
(%i5) lu_factor (A);
```

$$(\%o5)\quad [\begin{bmatrix} 5 & 4 & -2 \\ \frac{4}{5} & -\frac{1}{5} & \frac{28}{5} \\ -\frac{2}{5} & -28 & 162 \end{bmatrix}, [1,2,3], generalring]$$

CHAPTER 5

DIFFERENTIAL EQUATIONS

5.1 INTRODUCTION TO DIFFERENTIAL EQUATIONS

This section describes the functions available in Maxima to obtain analytic solutions for some specific types of first and second-order equations.

5.2 SOLVING ORDINARY DIFFERENTIAL EQUATIONS

ode2 and ic1

`gsoln : ode2 (de, u, t);`
`where de involves 'diff(u,t).`
`psoln : ic1 (gsoln, t = t0, u = u0);`

desolve

`gsoln : desolve(de, u(t) );`
`where de includes the equal sign (=)`
`and 'diff(u(t),t) and possibly u(t).`
`psoln : ratsubst(u0val,u(o),gsoln)`

plotdf

`plotdf ( dudt, [t,u], [trajectory_at, t0, u0],`
`[direction,forward], [t, tmin, tmax],`
`[u, umin, umax] )$`

rk
`points : rk ( dudt, u, u0, [t, t0, tlast, dt] )$`
`where dudt is a function of t and u which` `determines diff(u,t).`

METHODS FOR ONE FIRST ORDER ODE

EXACT SOLUTION WITH ODE2 AND IC1

Most ordinary differential equations have no known exact solution (or the exact solution is a complicated expression involving many terms with special functions) and one normally uses approximate methods. However, some ordinary differential equations have simple exact solutions, and many of these can be found using *ode2*, *desolve*, or *contrib ode*.

FUNCTION: ODE2 (EQN, DVAR, IVAR)

The function *ode2* solves an ordinary differential equation (ODE) of first or second order.

It takes three arguments:

- an ODE given by eqn,
- the dependent variable dvar,
- the independent variable ivar.

When successful, it returns either an explicit or implicit solution for the dependent variable.

%c is used to represent the integration constant in the case of first-order equations, and **%k1** and **%k2** the constants for second-order equations.

The dependence of the dependent variable on the independent variable does not have to be written explicitly, as in the case of *desolve*, but the independent variable must always be given as the third argument.

If the differential equation has the structure Left(dudt,u,t) = Right(dudt,u,t) (here u is the dependent variable and t is the independent variable), we can always rewrite that differential equation as de = Left(dudt,u,t) – Right(dudt,u,t) = 0, or de = 0.

We can use the syntax *ode2(de,u,t)*, with the first argument an expression which includes derivatives, instead of the complete equation including the " = 0" on the end, and ode2 will assume we mean de = 0 for the differential equation. (Of course, you can also use ode2 (de=0, u, t)).

Command	*Description*
`ode2(eqn,dvar,ivar)`	Solves ODE of first and second order
`ic1(sol,xval,yval)`	Solves initial value problems of first order
`ic2(sol,xval,yval,dval)`	Solves initial value problems of second order

Constant	*Description*
`%c`	integration constant for first order ODEs
`%k1, %k2`	integration constants for second order ODEs

System Variable	*Default*	*Description*
`method`		Shows successful solution method

An ordinary differential equation is an equation where the unknown is not a number but a function.

First-order differential equations contain first derivatives and can be written as

$$y' = f(x, y)$$

Second-order differential equations contain first and second derivatives:

$$y'' = f(x, y, y')$$

An example of a first-order derivative is

$$x^2 + \frac{dy}{dx} + 3yx = \frac{(\sin(\sin x)}{x}$$

There are two ways to represent ordinary differential equations in Maxima. The simplest way is to represent the derivatives by **'diff(y,x)** and **'diff(y,x,2)**, respectively. The above ordinary differential equation would then be entered as

```
(%i1) x^2*'diff(y,x) + 3*y*x = sin(x)/x;
```

$$(\%o1)\quad x^2\left(\frac{d}{d\,x}\,y\right)+3\,x\,y=\frac{\sin(x)}{x}$$

Note that the derivative **'diff(y,x)** is quoted, to prevent it from being evaluated (to 0). The second way is to write y(x) explicitly as a function of x. The above equation would then be entered as

```
(%i2) x^2*diff(y(x),x) + 3*y(x)*x = sin(x)/x;
```

$$(\%o2)\quad x^2\left(\frac{d}{d\,x}\,y(x)\right)+3\,x\,y(x)=\frac{\sin(x)}{x}$$

The function tries various integration methods to find an analytic solution. When successful, it returns either an explicit or implicit solution for the dependent variable. **%c** is used to represent the integration constant in the case of first-order equations. If routine *ode2* cannot obtain a solution for whatever reason, it returns false.

```
(%i3) ode2(x^2*'diff(y,x) + 3*y*x = sin(x)/x, y, x);
```

$$(\%o3)\quad y=\frac{\%c-\cos(x)}{x^3}$$

ode2 stores information about the method used to solve the differential equations in the variable **method**. The above differential equation has been recognized as a linear ODE.

```
(%i4) method;
(%o4) linear
```

In the case where y(x) is given, then the independent variable must be given as function **y(x)**.

```
(%i5) ode2(x^2*diff(y(x),x) + 3*y(x)*x = sin(x)/x, y(x), x);
```

(%o5) $y(x)=\frac{\%c-\cos(x)}{x^3}$

Initial value problems of first order can be solved by routine *ic1*. It takes a general solution to the equation as found by *ode2*as its first argument. The second argument *xval* gives an initial value for the independent in the form x=x0and the third argument *yval* gives the initial value for the dependent variable in the form y=y0.

We get the solution of the initial value problem

$$x^2 + \frac{dy}{dx} + 3yx = \frac{(\sin(\sin x)}{x}, y(\pi) = 0$$

in the following way:

```
(%i6) sol1: ode2(x^2*'diff(y,x) + 3*y*x = sin(x)/x, y, x);
```

(%o6) $y=\frac{\%c-\cos(x)}{x^3}$

```
(%i7) ic1(sol1, x=%pi, y=0);
```

(%o7) $y=-\frac{\cos(x)+1}{x^3}$

Similarly, we can use *ode2*to solve second-order differential equations. In this case, the two integration constants are represented by **%k1** and **%k2**.

```
(%i8) sol2: ode2('diff(y,x,2) + y*'diff(y,x)^3 = 0, y, x);
```

(%o8) $\frac{y^3+6\ \%k1\ y}{6}=x+\%k2$

The corresponding initial value problem can be solved by means of *ic2*. Compared to *ic1*is has an additional forth argument dval that gives the initial value of the first derivative of the dependent variable with respect to the independent variable in the form diff(y,x)=dy0.

```
(%i9) ic2(sol2, x=0, y=0, 'diff(y,x)=2);
```

$$(\%o9)\quad \frac{y^3+3y}{6}=x$$

5.3 LINEAR SYSTEMS

Command	*Description*
desolve(eqn, y)	Solves linear ODE
desolve([eqn_1,...,eqn_n], [y_1,..,y_n])	Solves system of linear ODEs
atvalue(expr, [x_1=a_1,...],c)	Assigns value **c** to **expr** at point **x=a**

Systems of linear equations can be found by routine *desolve*. It is important that the functions yi(x) must be given in the functional form. The solution of the linear second-order differential equation

$$y''(x)+y(x)=2x$$

can be found as follows:

```
(%i10) desolve(diff(y(x),x,2)+y(x)=2*x, y(x));
```

$$(\%o10)\quad y(x)=\sin(x)\left(\left.\frac{d}{dx}y(x)\right|_{x=0}-2\right)+y(0)\cos(x)+2x$$

If *desolve* cannot find a solution to the given ODE, then **false** is returned.

If initial conditions at x = 0 are known, they can be supplied by using *atvalue*. Notice that the conditions can only be given at 0, and must be given before the equations are solved. It is recommended to remove your assignment immediately after solving the initial value problem as otherwise, it may cause obscure messages or wrong results.

```
(%i12) atvalue(y(x), x=0, 1)$

(%i13) atvalue(diff(y(x),x), x=0, 2)$

(%i14) desolve(diff(y(x),x,2)+y(x)=2*x, y(x));
```

(%o14) $y(x)=\cos(x)+2\ x$

```
(%i15) kill(y)$
```

Similarly, systems of linear ODEs can be solved.

$$x'(t)=x(t)-2y(t)$$
$$y'(t)=-x(t)+y(t)$$

```
(%i16) eqn1: diff(x(t),t)=x(t)-2*y(t);
```

(%o16) $\frac{d}{d\,t}x(t)=x(t)-2\ y(t)$

```
(%i17) eqn2: diff(y(t),t)=-x(t)+y(t);
```

(%o17) $\frac{d}{d\,t}y(t)=y(t)-x(t)$

```
(%i18) desolve([eqn1,eqn2],[x(t),y(t)]);
```

(%o18) $[x(t)=\%e^{t}\left(\frac{(2\,(-2\ y(0)-x(0))+2\ x(0))\sinh(\sqrt{2}\ t)}{2^{3/2}}+x(0)\cosh(\sqrt{2}\ t)\right),y(t)=\%e^{t}\left(\frac{(2\ y(0)+2\,(-y(0)-x(0)))\sinh(\sqrt{2}\ t)}{2^{3/2}}+y(0)\cosh(\sqrt{2}\ t)\right)]$

The initial value problem with x(0) = 1 and y(0) = 1 is then solved by

```
(%i19)  atvalue(x(t), t=0, 1)$
       atvalue(y(t), t=0, 1)$
       desolve([eqn1,eqn2],[x(t),y(t)]);
(%i20)
```

(%o21) $[x(t)=\%e^{t}(\cosh(\sqrt{2}\ t)-\sqrt{2}\sinh(\sqrt{2}\ t)),y(t)=\%e^{t}\left(\cosh(\sqrt{2}\ t)-\frac{\sinh(\sqrt{2}\ t)}{\sqrt{2}}\right)]$

5.4 DIRECTION FIELDS

Command	*Description*
`plotdf(dxdy, [x,y],opts)`	Plot direction field of a single ODE
`plotdf([dxdt,dydt], [x,y],opts)`	Plot direction field of a set of two autonomous ODEs

Option	*Default*	*Description*
`[trajectory_at,x,y]`	empty	Starting point of an integral curve
`[direction,dir]`	`both`	direction of the independent variable to compute an integral curve (`forward`, `backward`, `both`)
`[versus_t,num]`	`0`	If `1`, curves are also plot as function of t
`[x,x_min,x_max]`	automatic	Range for x-axis
`[y,y_min,y_max]`	automatic	Range for y-axis

plotdf creates a plot of a vector field (direction field) of a first-order ODE or a system of two first-order ODEs. The directions field is a graphical representation of the solutions of an ODE.

DIRECTION FIELDS OF A SINGLE FIRST-ORDER ODE

To plot the direction field of a single ODE, the ODE must be written in the form

$$\frac{dy}{dx} = F(x, y)$$

and the function F should be given as the first argument for *plotdf*. The name of the independent and the dependent variable are given as a list as the second argument. If these are called **x** and **y**, respectively, it may be omitted.

When called *plotdf* opens a new window that contains the direction plot. The length of the draw vectors is proportional to the absolute value of the first derivative.

A direction field for the differential equation

$$y' = e^{-x} + y$$

is created by the following Maxima code. Notice that it might be necessary to load package *plotdf* first.

```
(%i1) load(plotdf)$
      plotdf(exp(-x)+y)$
```

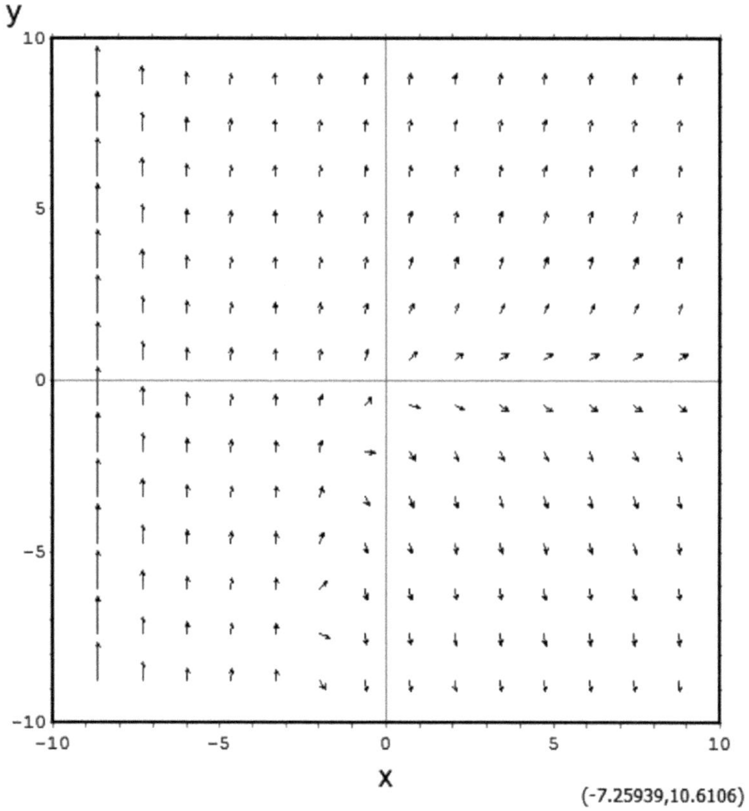

The menu in the plot window has several options (not shown in the figure as its appearance heavily depends on computing environment) that allows to manipulate the plot. Clicking into the direction field on a particular point adds a trajectory starting at that point.

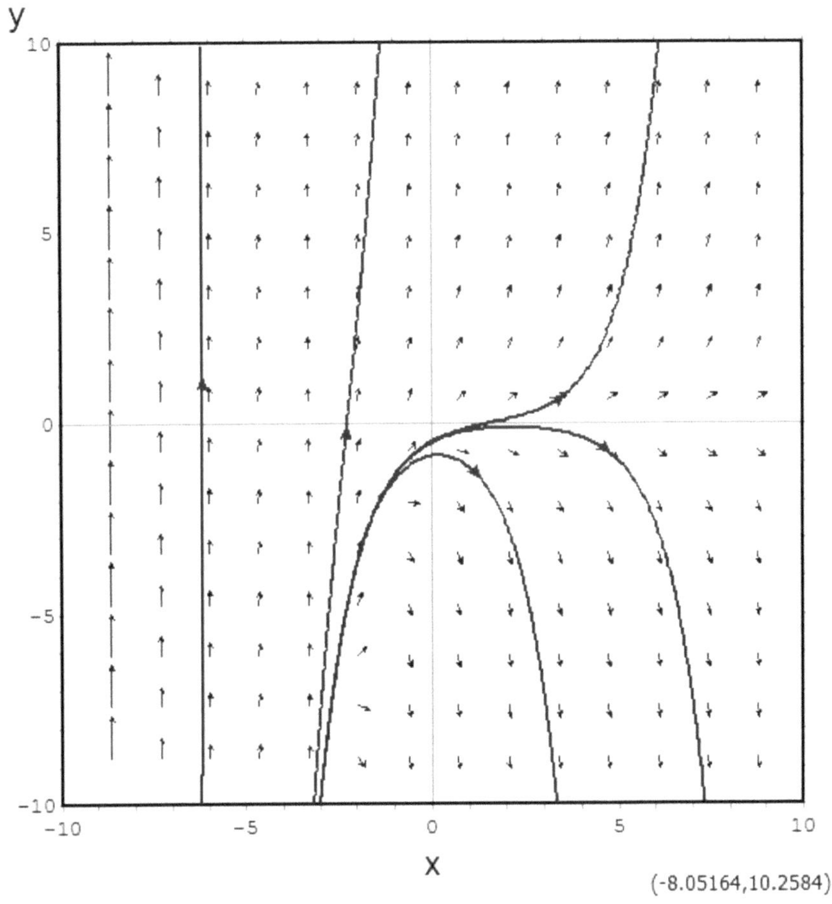

It is possible to draw two or more trajectories with different starting points. One can also enter staring points by means of option *trajectory_at*.

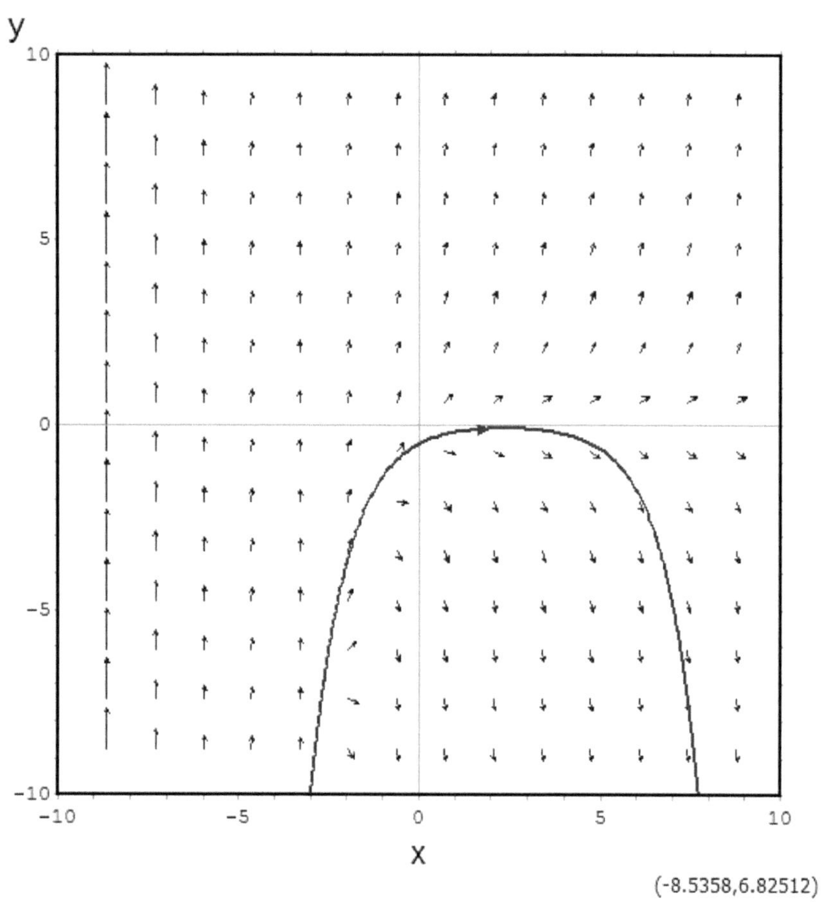

Option *direction* controls the direction of the independent variable that will be followed to compute an integral curve. Possible values are *forward*, to make the independent variable increase, *backward*, to make the independent variable decrease or *both* that will lead to an integral curve that extends in both directions.

It is also possible to plot the integral curve as a function of the independent variable t in a second plot window. This can be switched on using the appropriate option in the menu of the plot window or by means of command line option *versus_t*.

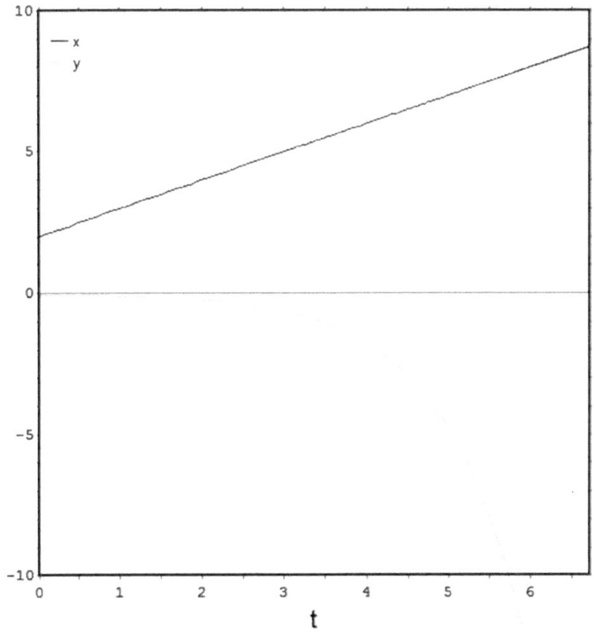

(See color insert.)

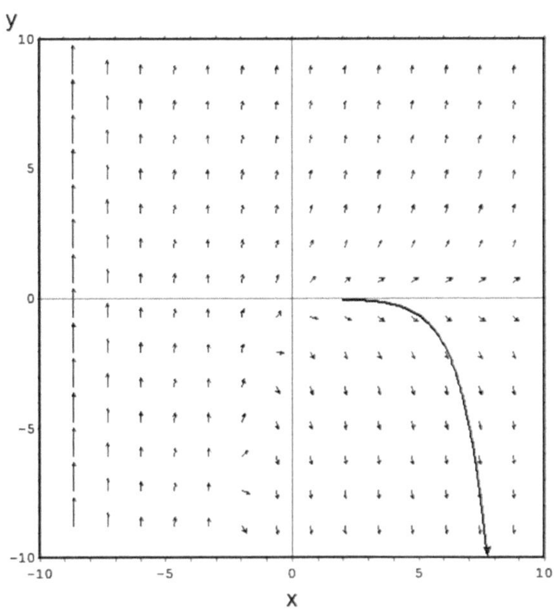

DIRECTION FIELDS OF TWO AUTONOMOUS FIRST-ORDER ODES

To plot the direction field of a set of two autonomous ODEs, they must be written in the form

$$\frac{dx}{dt} = F(x,y), \frac{dy}{dt} = G(x,y)$$

and the first argument for *plotdf* should be a list with the two functions F and G, in that order; namely, the first expression in the list will be taken to be the time derivative of the variable represented on the horizontal axis, and the second expression will be the time derivative of the variable represented on the vertical axis. Those two variables do not have to be x and y, but if they are not, then the second argument given to *plotdf* must be another list naming the two variables, first the one on the horizontal axis and then the one on the vertical axis.

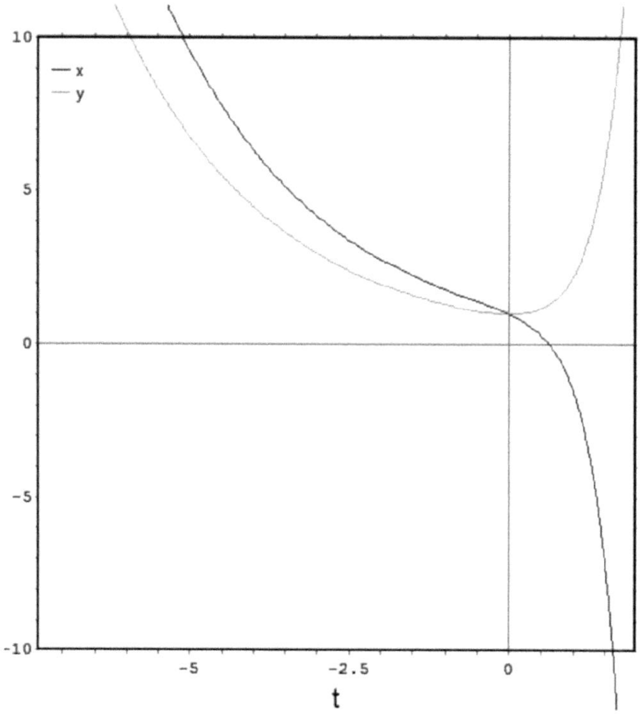

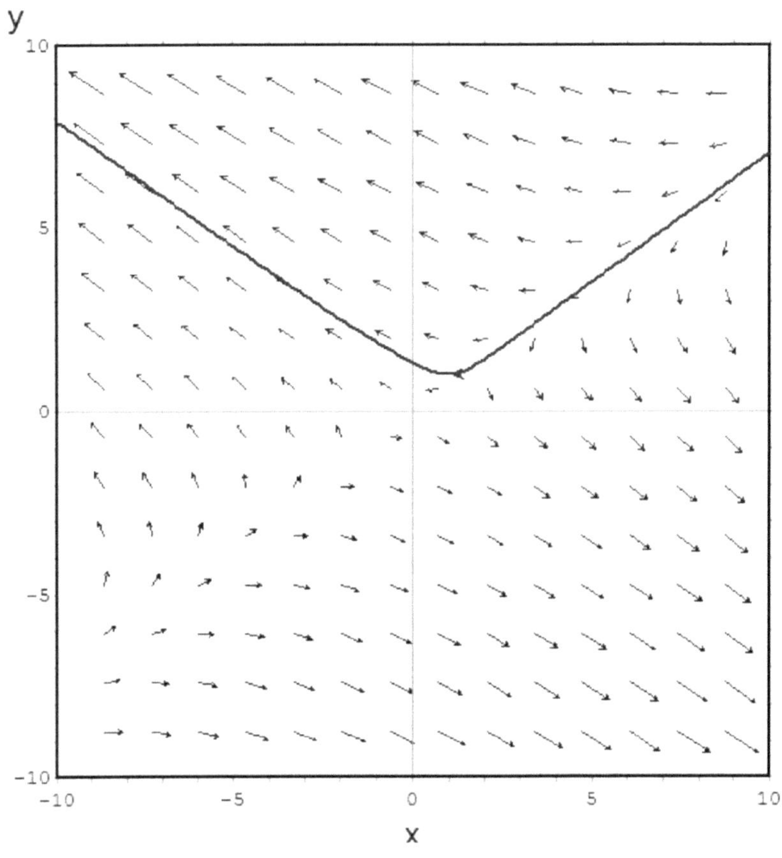

APPLICATION

1) Compute the general solutions of the following ordinary differential equations by means of routine **ode2** as well as the corresponding initial value problems with initial values y(1) = 1.
   Determine the methods that are used to solve these ODEs.

a) $y' - k\dfrac{y}{x} = 0$

```
(%i2) a: ode2('diff(y,x)-k*y/x=0,y,x);
```
(%o2) $y = \%c\ \%e^{k \log(x)}$

```
(%i3) ic1(a,x=1,y=1);
```
(%o3) $y = \%e^{k \log(x)}$

```
(%i4) method;
(%o4) linear
```

b) $xy' - (1+y) = 0$

```
(%i5) b: ode2(x*'diff(y,x)-(1+y)=0,y,x);
```
(%o5) $y = \left(\%c - \frac{1}{x}\right) x$

```
(%i6) ic1(b,x=1,y=1);
```
(%o6) $y = 2\ x - 1$

```
(%i7) method;
(%o7) linear
```

c) $y' = xy$

```
(%i9) c: ode2('diff(y,x)=x*y,y,x);
```

(%o9) $y = \%c\,\%e^{\frac{x^2}{2}}$

```
(%i10) ic1(c,x=1,y=1);
```

(%o10) $y = \%e^{\frac{x^2}{2}-\frac{1}{2}}$

```
(%i11) method;
(%o11) linear
```

d) $y' + e^y = 0$

```
(%i12) d: ode2('diff(y,x)+%e^y=0,y,x);
```

(%o12) $\%e^{-y} = x + \%c$

```
(%i13) ic1(d,x=1,y=1);
```

(%o13) $\%e^{-y} = \%e^{-1}(\%e\,x - \%e + 1)$

```
(%i14) method;
(%o14) separable
```

e) $y' = y^2$

```
(%i15) e: ode2('diff(y,x)=y^2,y,x);
```

$$(\%o15)\quad -\frac{1}{y}=x+\%c$$

```
(%i16) ic1(e,x=1,y=1);
```

$$(\%o16)\quad -\frac{1}{y}=x-2$$

```
(%i17) method;
(%o17) separable
```

f) $y' = \sqrt{x^3 y}$

```
(%i18) f: ode2('diff(y,x)=sqrt(x^3*y),y,x);
```

$$(\%o18)\quad -\frac{2\,x\sqrt{x^3\,y}-10\,y}{5\sqrt{y}}=\%c$$

```
(%i19) ic1(f,x=1,y=1);
```

$$(\%o19)\quad -\frac{2\,x\sqrt{x^3\,y}-10\,y}{5\sqrt{y}}=\frac{8}{5}$$

```
(%i20) method;
(%o20) exact
```

2) Compute the general solutions of the following ordinary differential equations by means of routine *ode2*as well as the corresponding initial value problems with the given initial values. Determine the methods that are used to solve these ODEs.

a) $y'' + y' - 2y = 3 \mathit{with}\ y(0) = y'(0) = 1$

```
(%i21) a: ode2('diff(y,x,2)+'diff(y,x)-2*y=3,y,x);
```
(%o21) $y=\%k1\ \%e^{x}+\%k2\ \%e^{-2\,x}-\frac{3}{2}$

```
(%i22) method;
(%o22) variationofparameters
```

```
(%i23) ic2(a,x=0,y=1,'diff(y,x)=1);
```
(%o23) $y=2\ \%e^{x}+\frac{\%e^{-2\,x}}{2}-\frac{3}{2}$

b) $y'' + 2y' + 17y = 0\ with\ y(0) = 0\ and\ y'(1) = 1$

```
(%i24) b: ode2('diff(y,x,2)-6*'diff(y,x)+9*y=0,y,x);
```
(%o24) $y=(\%k2\ x+\%k1)\ \%e^{3\,x}$

```
(%i25) method;
(%o25) constcoeff
```

```
(%i26) ic2(b,x=0,y=2,'diff(y,x)=0);
```
(%o26) $y=(2-6\,x)\ \%e^{3\,x}$

c) $y'' + 2y' + 17y = 0\ with\ y(0) = 0\ and\ y'(1) = 1$

```
(%i27) c: ode2('diff(y,x,2)+2*'diff(y,x)+17*y=0,y,x);
```
(%o27) $y=\%e^{-x}(\%k1\ \sin(4\,x)+\%k2\ \cos(4\,x))$

```
(%i28) method;
(%o28) constcoeff
```

```
(%i29) ic2(c,x=0,y=0,'diff(y,x)=1);
```
(%o29) $y=\frac{\%e^{-x}\ \sin(4\,x)}{4}$

3) Compute the general solutions of the following ordinary differential equations by means of routine *desolve* as well as the corresponding initial value problems with the given initial values.

a) $xy' - (1+y) = 0 \; with \; y(0) = 1$

```
(%i38) a: desolve(diff (y(x),x)-(1+y(x)), y(x));
```

(%o38) $y(x)=(y(0)+1)\,\%e^{x}-1$

```
(%i39) atvalue(y(x),x=0,1)$

(%i40) desolve(diff (y(x),x)-(1+y(x)), y(x));
```

(%o40) $y(x)=2\,\%e^{x}-1$

```
(%i41) kill(v)$
```

b) $y'' - 6y' + 9y = 3 \; with \; y(0) = 2 \; and \; y'(0) = 0$

```
(%i37) b: desolve(diff (y(x),x,2)+diff(y(x),x)-2*y(x)=3, y(x));
```

(%o37) $$y(x)=-\frac{\%e^{-2x}\left(2\left(\left.\frac{d}{d\,x}y(x)\right|_{x=0}\right)-2\,y(0)-3\right)}{6}+\frac{\%e^{x}\left(\left.\frac{d}{d\,x}y(x)\right|_{x=0}+2\,y(0)+3\right)}{3}-\frac{3}{2}$$

```
(%i33) atvalue(y(x),x=0,1);
(%o33) 1

(%i34) atvalue(diff(y(x),x),x=0,1);
(%o34) 1

(%i35) desolve(diff (y(x),x,2)+diff(y(x),x)-2*y(x)=3, y(x));
```

(%o35) $y(x)=2\,\%e^{x}+\frac{\%e^{-2x}}{2}-\frac{3}{2}$

```
(%i36) kill(y)$
```

c) $y'' + 2y' + 17y = 0 \;\; with\; y(0) = 0 \; and \; y'(1) = 1$

```
(%i82) c: desolve(diff(y(x),x,2)-6*diff(y(x),x)+9*y(x)=3, y(x));
       atvalue(y(x), x=0, 2); atvalue(diff(y(x),x), x=0, 0);
       desolve(diff(y(x),x,2)-6*diff(y(x),x)+9*y(x)=3, y(x));
       kill(y)$
```

(%o82) $y(x)=x\,\%e^{3\,x}\left(\left.\frac{d}{d\,x}\,y(x)\right|_{x=0}\right)-3\,y(0)\,x\,\%e^{3\,x}+x\,\%e^{3\,x}+\frac{(3\,y(0)-1)\,\%e^{3\,x}}{3}+\frac{1}{3}$

(%i83)

(%o83) 2

(%o84) 0

d) $y'' + 2y' + 17y = 0 \; with \; y(0) = 0 \; and \; y'(1) = 1$

```
(%i77) d: desolve(diff(y(x),x,2)+2*diff(y(x),x)+17*y(x)=3, y(x));
       atvalue(y(x), x=0, 0); atvalue(diff(y(x),x), x=0, 1);
       desolve(diff(y(x),x,2)+2*diff(y(x),x)+17*y(x)=3, y(x));
       kill(y)$
```

(%o77) $y(x)=\%e^{-x}\left(\frac{\sin(4\,x)\left(\frac{2\left(17\left(\left.\frac{d}{d\,x}\,y(x)\right|_{x=0}\right)+34\,y(0)-6\right)}{17}-\frac{2\,(17\,y(0)-3)}{17}\right)}{8}+\frac{(17\,y(0)-3)\cos(4\,x)}{17}\right)+\frac{3}{17}$

(%i78)

(%o78) 0

(%o79) 1

(%o80) $y(x)=\%e^{-x}\left(\frac{7\,\sin(4\,x)}{34}-\frac{3\,\cos(4\,x)}{17}\right)+\frac{3}{17}$

(%i81)

4) Plot the direction fields and some trajectories for the following ordinary differential equations.

a) $y' - k\dfrac{y}{x} = 0$

```
(%i2) plotdf(y/x);
```

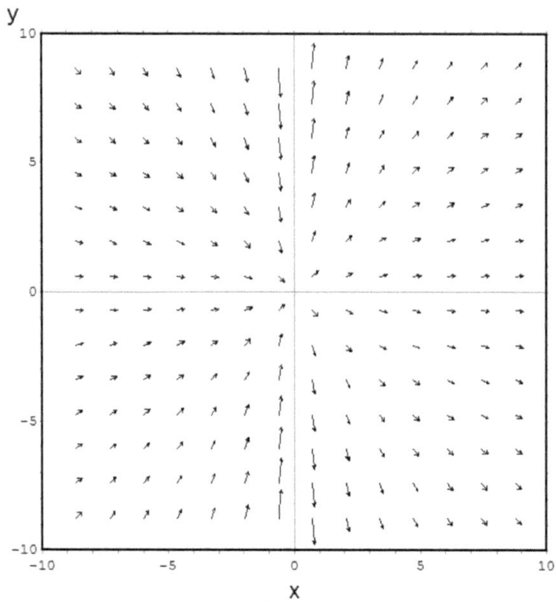
y
10
5
0
-5
-10
-10
-5
0
5
10
x

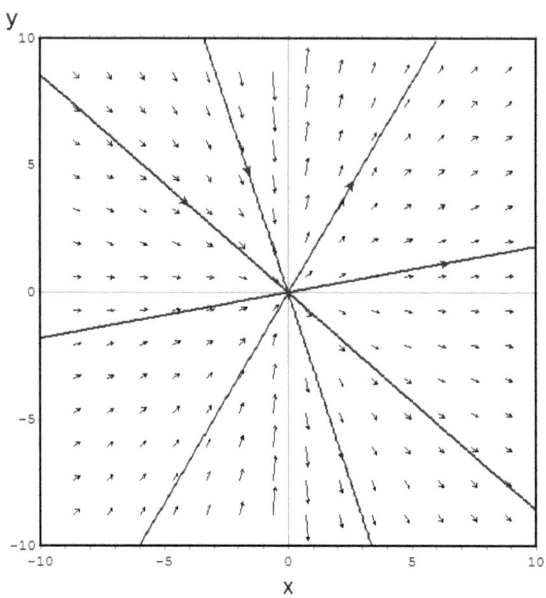
y
10
5
0
-5
-10
-10
-5
0
5
10
x

b) $xy' - (1 + y) = 0$

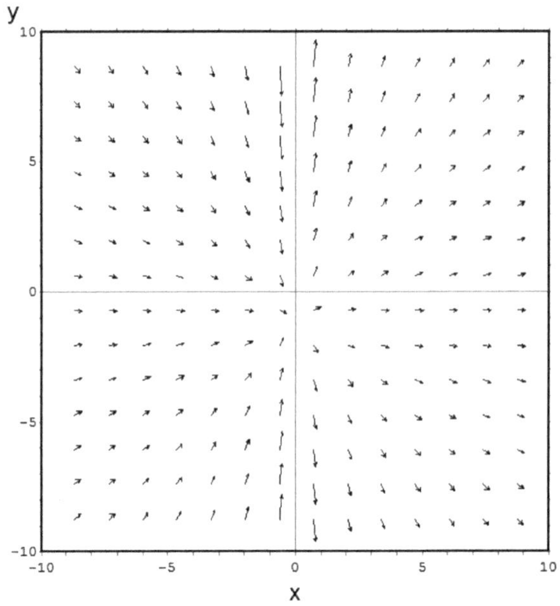


```
(%i4) plotdf((1+y)/x);
```

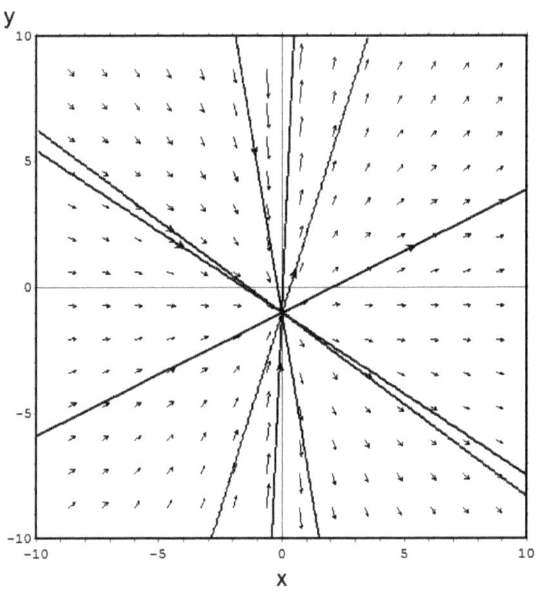

c) $y' = xy$

```
(%i1) plotdf(y*x);
```

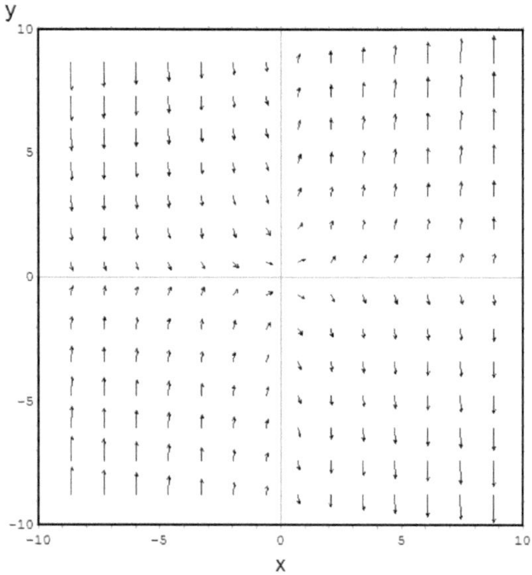

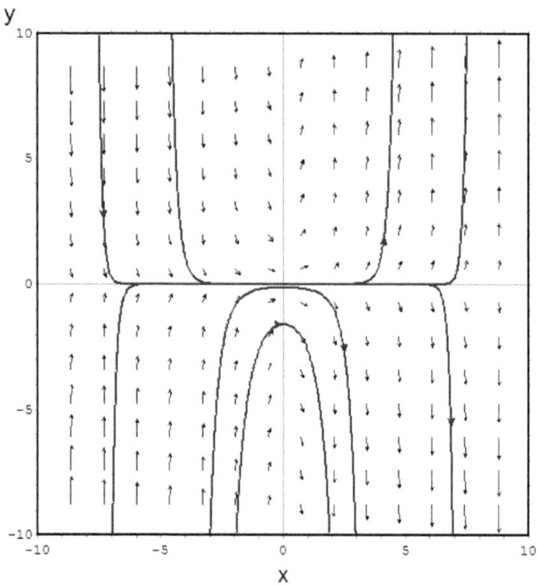

d) $y' + e^{y} = 0$

```
(%i2) plotdf(-%e^y);
```

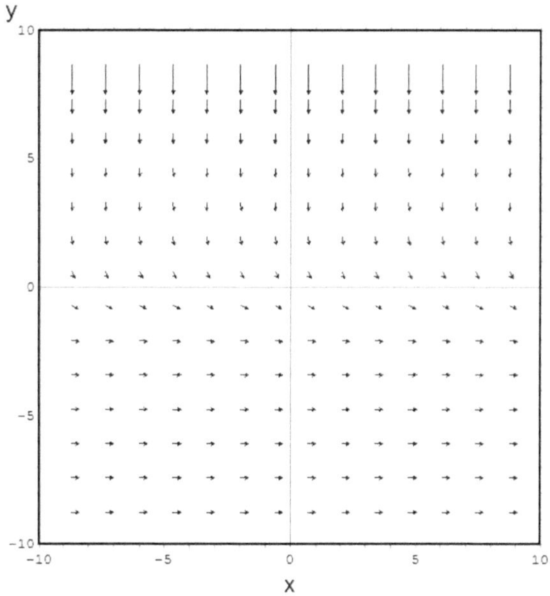

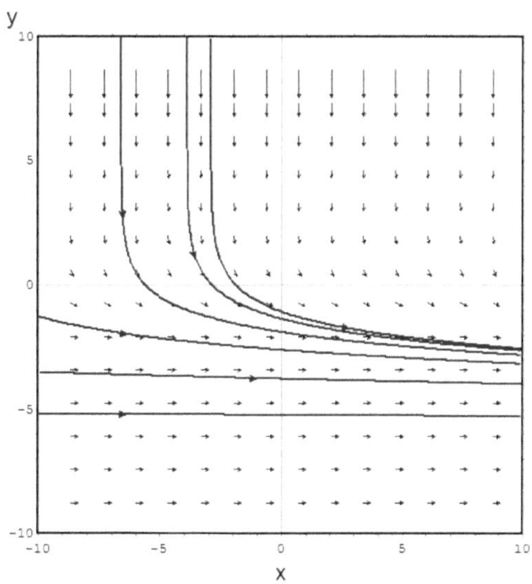

e) $y' = y^2$

```
(%i4) plotdf(y^2);
```

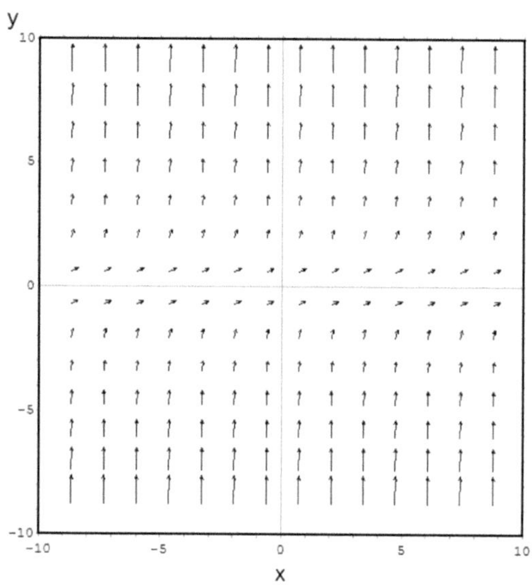

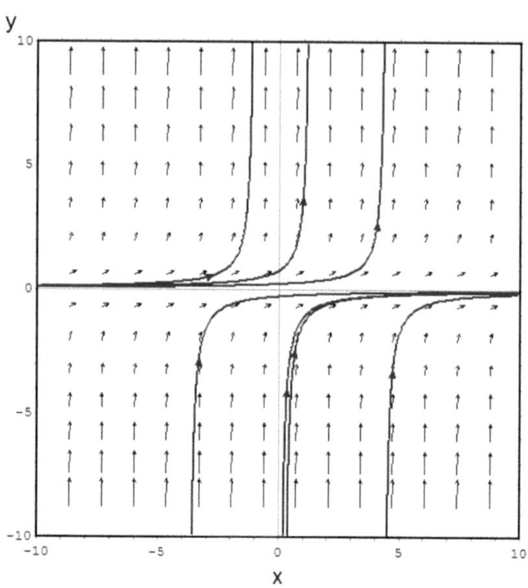

5) Plot the direction fields and some trajectories for the following system of ordinary differential equations. Use plotting range [0, 10] × [0, 10].

$$x'(t)-x(t)\big(1-y(t)\big)$$
$$y'(t)=-y(t)\big(1-x(t)\big)$$

```
(%i10) plotdf([x*(1-y),-y*(1-x)],[x,0,3],[y,0,3]);
```

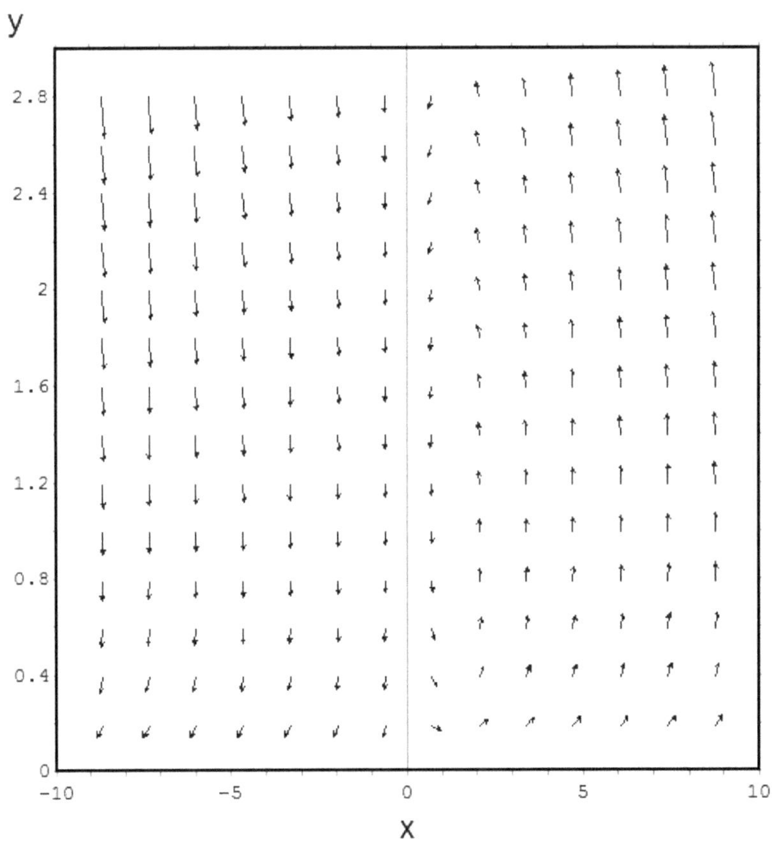

INDEX